AF830554

NANO INTERCONNECTS

NANO INTERCONNECTS

Device Physics, Modeling and Simulation

Afreen Khursheed and Kavita Khare

CRC Press
Taylor & Francis Group
Boca Raton London New York

CRC Press is an imprint of the
Taylor & Francis Group, an **informa** business

First edition published 2022
by CRC Press
6000 Broken Sound Parkway NW, Suite 300, Boca Raton, FL 33487-2742

and by CRC Press
2 Park Square, Milton Park, Abingdon, Oxon, OX14 4RN

© 2022 Afreen Khursheed and Kavita Khare

First edition published by CRC Press 2022

CRC Press is an imprint of Taylor & Francis Group, LLC

Reasonable efforts have been made to publish reliable data and information, but the author and publisher cannot assume responsibility for the validity of all materials or the consequences of their use. The authors and publishers have attempted to trace the copyright holders of all material reproduced in this publication and apologize to copyright holders if permission to publish in this form has not been obtained. If any copyright material has not been acknowledged please write and let us know so we may rectify in any future reprint.

Except as permitted under U.S. Copyright Law, no part of this book may be reprinted, reproduced, transmitted, or utilized in any form by any electronic, mechanical, or other means, now known or hereafter invented, including photocopying, microfilming, and recording, or in any information storage or retrieval system, without written permission from the publishers.

For permission to photocopy or use material electronically from this work, access www.copyright.com or contact the Copyright Clearance Center, Inc. (CCC), 222 Rosewood Drive, Danvers, MA 01923, 978-750-8400. For works that are not available on CCC please contact mpkbookspermissions@tandf.co.uk

Trademark notice: Product or corporate names may be trademarks or registered trademarks and are used only for identification and explanation without intent to infringe.

Library of Congress Cataloging-in-Publication Data
A catalog record has been requested for this book

ISBN: 978-0-367-61048-7 (hbk)
ISBN: 978-0-367-61115-6 (pbk)
ISBN: 978-1-003-10419-3 (ebk)

DOI: 10.1201/9781003104193

Typeset in Times
by MPS Limited, Dehradun

Contents

List of Figures

List of Tables

Preface

From the invention of the microprocessor unit (MPU) in 1971, the VLSI industry has made noteworthy strides in enhancing the on-chip circuit performance by improving interconnect latency, reducing the power and packing more blocks into integrated circuits (ICs). All of these advancements were mainly governed by systematic transistor scaling, which quadrupled the IC performance with the upcoming technology generation. The performance of the primitive MPU was mainly restrained by gate transistor speeds, and the impact of wire interconnects on the system performance was insignificant. However, with further downsizing of technology, the transistor performance enhances in leaps and bounds, albeit the interconnect wire performance considerably deteriorated due to the surge resistance of the wires with smaller dimensions. This pushes the semiconductor industry to make a one-time move from conventional to new-fangled interconnect conductor material. Despite of this, it is observed that an average of 50% of the energy dissipated in MPU was consumed solely by wire interconnects, and almost half of the interconnect power absorbed by short local interconnects only. In addition to this, with the adoption of innovative device technologies like strain-enhanced MOSFETs and FINFETs as to mention a few, to enrich the device performance further aggravates the interconnect issue. As a result, many state-of-the-art on-chip interconnect technologies like optical interconnects, plasmonics interconnects, spintronics interconnects, carbon nanotubes and graphene nanoribbon interconnects are being explored.

This book ferrets out the issues encountered by VLSI circuits due to scaling of on-chip interconnect dimensions. The primary goal of this manuscript is to bring forth a comprehensive analysis of ultra-low-power, high-speed nano interconnects based on different facets like material modeling, circuit modeling and the adoption of repeater insertion strategies and measurement techniques pertinent to current research scenarios.

As "brevity is the soul of wit," hence we restrict ourselves with the analysis of specifically local interconnect and semi-global or intermediate models that are incorporated between two circuits on a single chip rather than globally connected chip-to-chip interconnects. This transcript will be instrumental to the post-graduate students and researchers working in the area of low-power VLSI circuits by giving a deep insight into the carbon allotropes (CNT and GNR) based VLSI interconnects at nanometric regime.

The book is comprised of four chapters.

Chapter 1: This chapter gives an overview of the difficulties encountered in terms of interconnect performance deterioration with the advancement in VLSI industries and technology scaling. It also presents an extensive literature review on the background of interconnect evolution, their growing trend with upcoming future technologies and different strategies adopted by various researchers to handle the existing problem. However, just a glance at this introductory chapter will be sufficient for well-informed readers.

Chapter 2: It gives an introduction on types of interconnects highlighting their hierarchical structure in ASIC. It emphasizes the use of graphene (CNT and GNR) based interconnect as a substitute to conventional interconnect materials. In the end, various existing models of interconnect structures are presented and descriptions about their electrical behavior is also given.

Chapter 3: Repeater buffer insertion is a trusted approach for reducing delay and enhancing the performance of on-chip interconnects. But their exponentially increase in number for very long wire lines is becoming a hurdle in producing energy-efficient ICs. This chapter initially provides a background of traditional repeater interpolation methodology implemented in interconnects' network to mitigate propagation delay. It then mentions the various styles of buffer insertion to overcome the limitation of area overhead and power consumption. It further suggests the use of state-of-the-art device modeling to realize high-speed ultra-low-power smart buffers. The structure of novel CNTFET/GNRFET-based buffers and their compatibility with the existing as well as emerging wire technology and device technology is also explained in the end.

Chapter 4: The final chapter concludes with its first section presenting the brief discussion on signal integrity challenges encountered when modeling state-of-the-art nano interconnects. As feature size decreases due to scaling of device dimensions, the reliability is compromised. The decrease in reliability is due to electromigration-induced problems. At a very high frequency of operation, the tightly packed nano interconnects produce transient crosstalk. Furthermore, analysis on how the crosstalk noise strongly influences the signal propagation delay and causes the circuit malfunction or functional failure is done. As the crosstalk noise causes signal overshoot, undershoot and ringing effects, hence it is enviable to develop a perfect pedagogical model for analytically analyzing the crosstalk effects occurring in nano interconnects.

For this reason, the later part of this chapter presents a precise and time-efficient model of FET-gate-driven coupled (copper/CNT/GNR) nano interconnects in order to commensurate the crosstalk incited performance analysis of nano interconnects. The model discussed is formulated by finite difference time domain (FDTD) methodology for the driver interconnect load (DIL) system by assimilating boundary conditions for all the three types of interconnect materials. The FET driver is modeled by either the nth power law model by considering the finite drain conductance parameter.

Acknowledgments

Writing a book on device physics of emerging technology-based novel interconnects is a bit more strenuous and assiduous than we thought; yet more guerdoning and appeasing than we could have ever imagined. Henceforth, at the onset we send our gratitude and praise to *Almighty* for His divine motivation bestowed on both of us to remain steadfast and vanquish several arduous phases during the preparation of this book.

We are very grateful to Dr. Gagandeep Singh, Mr. Gauravjeet Singh Reen and Mr. Lakshay Gaba of CRC Press, Taylor & Francis Group for the exceptional support and extended cooperation from their editorial office to pursue this book project.

We also acknowledge the love and care we received during these months from our families. It is only because of their perseverance and patience that we could complete this work in a timely manner.

Our sincere thanks go to our colleagues and friends for all their support that was needed for our work.

Last but not the least, the authors take this opportunity to thank all of our reviewers who devoted their valuable time in improving the final typescript. We incorporated most of your feedback and apologize that the limited time prevented some feedback from coming through. We sincerely apologize to the readers for any unintentional mistakes that may have crept into this book. All suggestions and feedback for further improvement of the book are welcome.

About the Authors

Dr Afreen Khursheed is currently an assistant professor in IIIT Bhopal. She earned a B.E. (EC) with honors in 2005 and M.Tech with honors in 2009 in VLSI and EMBEDDED SYSTEM from Maulana Azad National Institute of Technology, Bhopal. She earned her doctorate in nanotechnology from M.A.N.I.T, Bhopal.

Dr Khursheed has 13 years of teaching and research experience, and has successfully guided a M.Tech thesis. To her credit, Dr Afreen has one patent granted and one published patent which is in process of grant; she has publications in various international conferences and SCI (Q1/Q2) indexed journals, and is co-author of two books on low-power VLSI Design. Dr Afreen has also reviewed manuscripts for international journals and also served as the editorial board member of UGC journals and conferences. She has presented posters and also delivered lectures in various national and international events and conferences held at M.A.N.I.T, Texas A and M University, Qatar university, among others. Her areas of interest includes high-speed nanoelectronic circuit modeling for portable devices and ultra-low-power sub-threshold circuit simulation and device modeling using CNT and GNR. Her current area of interdisciplinary research focuses on VLSI implementation on a real-time operating system using spintronics.

Dr. Kavita Khare earned her B.Tech degree in electronics and communication engineering in 1989, M.Tech. degree in digital communication systems in 1993 and Ph.D. degree in the field of VLSI design in 2004. She has nearly 29 years teaching experience. She has 200 publications in various international conferences and journals like IEEE, springer, Elsivier, Oxford press, Hindawi, European Transactions on Telecommunications, Inderscience publications., Taylor & Fancis, CSIR (NISCAIR), AMSE. Dr Khare has guided 17 PhD theses and 40 Mtech theses. She has a best paper award from an International Conference in France, one patent granted and two are published. She is coordinator of short-term courses on VLSI design organized at MANIT Bhopal. She is a reviewer of papers of IEEE Jour. of Circuits & Systems, Microelectronics Journal, Elsevier, International Journal AMSE (France), Institute of Engineers and many international conferences. She is the editor/ chief advisory board member of some international journals.

She is currently a professor and head of the Department of Electronics and Communication Engineering in MANIT, Bhopal, INDIA. Her fields of interest are VLSI design and communication systems. Her research mainly includes design of arithmetic circuits and various communication algorithms related to synchronization, estimation and routing. She is a fellow IETE (India) and member of IEEE and life member ISTE.

About the Author

1 Prefatory Concepts and More

1.1 INTRODUCTION

The science behind the nanoscale materials, interconnect wires and devices along with their adoption in nanocomputers, ASIC and sensors establishes the sphere of nanotechnology. The prefix term "nano" denotes a basic unit on a length scale of the order of 10^{-9} meters. At this size, the circuits and systems are scaled down to reach the limit of 10–100 s of an atom where there are device physics and chemistry changes. It forms the basis for the progeny of avant-garde products based on the ultimate miniaturization where extended atomic or molecular structures form the basic building block.

The More Moore International Focus Team (IFT) has forecasted in the International Road map of Devices and Systems (IRDS) 2017 edition that downsizing of integrated circuits in accordance with Moore's law will result in several performance issues in the coming future (Bernd, 2012). Stepping into the deep nanometer era leads to the brink of implementing entire systems on a single chip. As a result of excessive scaling done to meet this requirement, a non-fortuitous obstacle has cropped up in the form of interconnect wire structures (Bohr, 1995). At the nanometer regime, as illustrated in Figure 1.1, the interconnect wire delays become more discernible than transistor gate delays and affect the performance of ultra-high frequency integrated circuits (International Technology Roadmap for Semiconductors (ITRS) Reports, 2013).

The crossover point depicts the start of the "interconnect bottleneck," where the repeater buffer insertion technique is adjunct to adoption of nano materials technology for interconnect wire designing could reduce this problem.

1.1.1 Prolegomenon to Material for Interconnects Designs

Interconnects does not merely behave as a simple resistor at higher frequencies but also has parasitic impedances concomitant with them (Ho et al., 2001). These impedance parameters are material dependent and become a bottleneck, thereby masking the performance of ICs in terms of the signal propagation delay, crosstalk noise, thermal stability, power dissipation and speed of circuits (Rabaey et al., 2003). During the outset of the chip industry, the decision regarding the choice of appropriate material, suitable for fabricating interconnect wire structures, was mainly determined by the ease of deposition, conductivity of material, high melting point and its affinity and adherence to SiO_2. Aluminium, which suffices these requirements, was initially the preferred interconnect material for providing electrical connection among the nodes in an IC (Weste et al., 2008), but an exponential

DOI: 10.1201/9781003104193-1

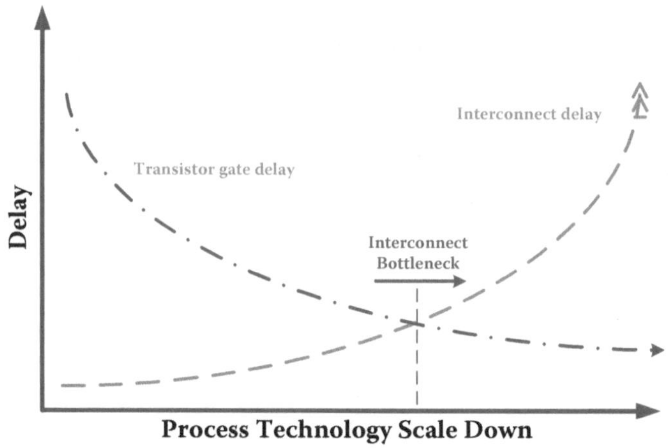

FIGURE 1.1 Trends in transistor gate latency (switching time) and interconnect latency (signal propagation delay) with technology scaling.

increase in device density due to advent of nanometre technology results in surge of current density. The limitation faced by aluminium is that for high current densities the problem of electromigration creeps in and hence Al was repudiated and replaced by copper. To date, copper is unanimously used as interconnect material in industries (Khursheed et al., 2018)Khursheed 2018. Further technology advancement leads to continuous scaling of semiconductor devices as a result of which even the Cu gradually started facing the problem of an increase in resistivity due to surface roughness and grain boundary scattering (Ceyhan and Naeemi, 2013). Moreover, at the nanometer range, the current density of interconnect wire exceeds the maximum limit of current that can be carried by copper conductors (6×10^6 A/cm^2) due to electromigration. This necessitates the requirement of material innovation. The conventionally used metallic conductor material must be replaced by carbon allotropes such as graphene as a new interconnect material. Graphene is a zero-bandgap semiconductor that shows metallic behavior because its conduction and valence bands meet at the Dirac point on the edge of the Brillouin zone. Graphene-based nano interconnects can be sorted as carbon nanotubes (CNTs) and graphene nanoribbons (GNRs). Both variants of graphene can carry up to $\approx 10^9$ A/cm^2 of maximum current density without fail and thus prove to be viable materials for interconnects (Khursheed et al., 2018).

1.1.2 Prolegomenon to Interconnects Repeater Buffer Design

In addition to using suitable interconnect material, further improvements in the performance of on-chip interconnects can be obtained by ratifying the ubiquitous strategy of repeater insertion. Repeaters are basically conventional inverters or modified buffer amplifiers inserted in between lengthy interconnect wires to tinker the signal degradation caused by parasitic impedance of interconnect wires. These

RLC parasitic impedances increase the signal propagation delay and power dissipation besides introducing extra noise source and thus affecting the reliability of circuit. The Elmore delay of any unbuffered interconnect wire is proportional to the square of the wire length. By using the technique of decimating a long interconnect wire into smaller subsections and inserting buffers as repeaters between the wires, the time delay dependency as the quadratic function of length will transform to a linear function of length. Plying of long interconnect wires not only engenders inordinate signal propagation delay but also adds to significantly large coupling capacitance and mutual inductance among the neighboring nets. At the deep submicron range due to shrinking of line width and spacing between the wires, the coupling capacitance and mutual inductance becomes instrumental in making the propagating signal more susceptible to crosstalk noise (Das and Rahaman, 2011). Thus, insertion of buffers at proper intervals simply breaks the lengthy wires into small sections and hence suppresses the crosstalk noise.

Interpolation of buffers as repeaters seems to be an effective method for mitigating crosstalk noise apart from reducing delay and regenerating the deformed output waveforms. However, the ramification of putting in too many repeaters causes switching power dissipation. Conventional buffers also fritter leakage power during an idle state. A peculiar trend has been observed that initially overall circuit delay decreases with the insertion of repeaters but after a requisite number of repeaters the delay increases (Pandya et al., 2012). This is so because the repeater itself exhibits a switching delay, which contributes to overall system delay. Hence, an optimum number must be decided to get a better result.

Thus, all of these factors into consideration, it can be said that there is an imperative need to design smart buffers as repeaters, which helps in enhancing interconnect performance by improving the speed of interconnect structures without affecting the dynamic as well as static power saving. Smart buffers are designed using novel nano materials, keeping in mind that these repeaters are ultra-low-power drivers of minimum size, which can discharge the load capacitance at a high speed in order to reduce delay.

1.2 EPOCH OF ON-CHIP INTERCONNECTS

1.2.1 Introduction

From the invention of the microprocessor unit (MPU) in 1971, the VLSI industry has made noteworthy strides in enhancing the on-chip circuit performance by improving interconnect latency, reducing the power and packing more blocks into integrated circuits (ICs). All of these advancements were mainly governed by systematic transistor scaling, which quadrupled the IC performance with the upcoming technology generation. The performance of the primitive MPU was mainly restrained by gate transistor speeds, and the impact of wire interconnects on the system performance was insignificant. However, with further downsizing of technology, the transistor performance enhances in leaps and bounds, albeit the interconnect wire performance considerably deteriorated due to the surge resistance of the wires with smaller dimensions. This pushes the semiconductor industry to make a one-time move from conventional to newfangled interconnect conductor material. Despite this, it is observed that on an average, 50% of the energy dissipated in the MPU was consumed solely by wire interconnects, and almost half of the interconnect power was absorbed by short local interconnects only. In addition to this, with the adoption of innovative device technologies like strain-enhanced MOSFETs and FINFET to mention a few, to enrich the device performance further aggravate interconnect issue. As a result, many state-of-the-art on-chip interconnect technologies like optical interconnects, plasmonics interconnects, spintronics interconnects, carbon nanotube and graphene nanoribbon interconnects are being explored.

The succeeding section presents a brief review on the extensive research work done till date in the domain of ultra deep submicron interconnects at various technology nodes. It also throws light on the background of on-chip interconnect modeling and the evolution of novel interconnects technology. Furthermore, it scrutinizes the various strategies adopted by the researchers at different stages of VLSI design flow to overcome the limitations encountered in the process of enriching the interconnect performance of an ASIC.

1.2.2 Background and Evolution of Interconnect Technology

Ceyhan and Naeemi together have carried out an extensive research based on the road map presented by ITRS and predicted that as the traditional interconnect technology of Cu is at the brim of scaling limits, mainly because of dwindling grain size with the shrinking of feature dimensions. All these factors lead to a hike in conductor resistivity and grain boundary scattering. They have pointed out the limitations of Cu and suggested that at the nanoscale regime the nature of the interconnect wire problem alters and paves the way for novel opportunities. Their work emphasizes that historical trend of attaining minimum interconnect latency at local and intermediate interconnects level will not hold true for upcoming future technology nodes. They also explored innovative opportunities that arise as a result of this radical change in the trend of the interconnect wire problem. Consequently, materials like quantum wires known as nanotubes, with large mean free path (MFP) up to few micrometers at room temperatures, huge charge carrying capacity, resistance to electromigration and excellent thermal conductivity are needed to meet the future challenges given by ITRS 2009 (Ceyhan and Naeemi, 2013).

Steinhogl, along with his team, conducted a comprehensive study of Cu/Low-k interconnects technologies for sub-100 nm CMOS ICs. To perform a systematic investigation different samples having width ranging from 40 nm to 1,000 nm and of distinct heights were surveyed. The conductor wires of copper were prepared in a silicon oxide matrix and electrically characterized. The temperature coefficient of resistance was considered to extract resistivity of obtained samples. To analyze the electrical data, models based on physical parameters were taken to further the sensitivity of model parameters extracted by best-fit procedure are investigated. It was reported that the impact of width and height on the resistivity, the influence of electron scattering at grain boundaries compared to surface scattering and the impact of grain sizes and impurities all are imposing a strong influence on the overall performance of the system through increased crosstalk, signal transmission delay and power dissipation. Nevertheless, Cu/Low-k interconnects have confronted electrical issues as well as material issues up to a great extent at the nanometer regime (Steinhogl et al., 2005).

Pawan Kapur et al. developed a model for the future technologies of copper interconnect resistivity. It is again verified by them on the basis of their model that dimensional scaling increases resistivity of copper wires, due to the impact of diffusion barrier and surface scattering. Apart from this, they also carried out a comparative analysis on the resistance of aluminium interconnects to copper wire at a 35 nm technology node. The result of analysis reveals the per unit length (p.u.l.) increase in resistance of 192%, 145% and 90% for local, semi-global and global lengths, respectively. Although the model developed by them is an ice-breaker for on-chip interconnects, the limitation with the readily adoption of this analysis for future studies at lower technology nodes is that they have not considered and in fact completely overlooked the effect of electron scattering through grain boundaries (Kapur et al., 2002).

Banerjee and Srivastava (2006) investigated for the ITRS intermediate wires for temperatures around 300 K that dimension scaling increases the resistivity of metal, grain boundary and surface scattering of electrons. The influence is even greater for local wires bearing the smallest cross-sectional area. It is inferred that Cu wires with dimensions of the order of MFP at room temperature exhibit low conductance in forthcoming technologies. As illustrated from Figure 1.2, the resistivity alters at 300 K, due to enhanced grain boundary scattering, surface scattering and the presence of a highly resistive diffusion barrier layer at local, semi-global and global level. The ill effect of a surge in resistivity is a hindrance in interconnect latency and also restricts the maximum charge-carrying capability of future Cu conductors. Their simulation results reveal that the use of carbon nanotubes for on-chip interconnects applications can significantly lower intermediate-level wire delay by more than 60% because of its smaller value of resistance, in spite of imperfect metal nanotube contacts. From the

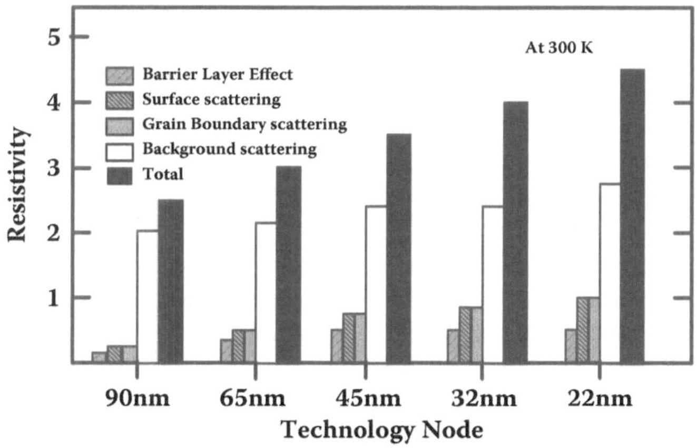

FIGURE 1.2 Effect of scaling on Cu resistivity for the ITRS intermediate wires at 300 K.

comparative investigations, it can be inferred that the suitability of CNTs for imminent on-chip interconnects is primarily due to the outstanding intrinsic properties of metallic single-walled CNTs adjunct with their remarkable performance, power and thermal reliability. Yet many challenges still need to be overcome in the areas of fabrication and process integration, although it is expected that these challenges will not cause any fundamental constraints for interconnect applications. Mitigating metal nanotube contact resistance is required for a local interconnect. Apart from that, thorough modeling of electromagnetic interactions and characterization of CNTs, 3D (metal) to 1D (CNT) contact resistance, impact of defects on electrical and thermal transport and high-frequency effects need to be quantified.

Srivastava et al. explained that most of the research work undertaken mainly focuses on investigating the CNT interconnects from perspective of performance only. The study carried out by them is an attempt to bridge the gap found in previous researches. For this purpose, a comprehensive evaluation of conventional Cu with CNT interconnects is executed. Quantification of performance metrics in terms of power consumption and reliability is done while considering all the practical constraints of nanotechnology. Based on the interconnect structure of a single-walled carbon nanotube (SWCNT), CNT bundle formulation of electrical circuit parameters is done (Srivastava et al., 2005).

The results justify that with the use of CNTs, an enhancement in the performance of lengthy global wires is attained by 80% with minimal additional power dissipation for maximum metallic CNT density. Investigations are further done by applying the power-optimal repeater insertion technique for both CNT and Cu interconnects. This approach maintains a trade-off in between minimum delay in conjunction to reduced power. Astonishingly, from the findings, it is vivid that a reduction in the rise of wire temperature is attained with carbon tubes; henceforth, it is suggested that if they are integrated with Cu wires then an improvement in performance of about 30% can be achieved. CNT is propitious material in tuning the back-end temperature and have significant implications for upcoming technologies, such as 3D ICs where thermal management is a major concern.

Murugeswari et al. (2015) demonstrated a comparative performance analysis of carbon-generated SWCNT and graphene nanoribbons against traditional Cu interconnects. The finding of the analysis justifies the replacement of Cu with newfangled carbon-allotrope-based wire material for on-chip interconnects. The results explicitly highlight that CNT structures offer more bandwidth and have less delay when compared to Cu. It is noted that due to their ballistic transport and excellent thermal and electrical, SLGNR and SWCNT generate only 0.5% and 0.7% of copper delay in the global interconnect level and have 1,000 times better energy efficiency than copper and also offer less power dissipation.

Li et al. (2009) scrutinized both carbon nanotubes and graphene nanoribbons generated interconnects on the basis of modeling perspectives and fabrication techniques and then presented a thermal and electrical model for both of the interconnects. A comparative result demonstrates the inferiority of Cu with respect to SWCNT and MWCNT global interconnects in terms of propagation delay. MWCNT vias are explicitly metallic; hence, suitable for sculpting wires. Electro thermal analysis of SWCNT vias justifies their suitability for TSVs in various chip-to-packaging interconnect applications as well as for 3D IC integration. Further study shows that taller CNT vias offers better advantages over the shorter ones and due to the presence of large kinetic inductance, the skin effects are mitigated, making CNTs as a promising prospect for future high-frequency circuit applications, whereas, in order to make GNRs comparable to copper or CNT wires, intercalation doping with high edge specularity is required.

Li et al.'s (2006) research work undertaken by their team has developed a model for metallic nanotubes, creating an analogy with that of a transmission line model to calculate circuit parameters in order to study their electrical behavior. The model is developed on the assumption that fundamental resistance or quantum resistance is considered a lumped parameter, whereas the scattering resistance is regarded as a distributed parameter. From their observations, they have suggested that the CNT bundle structure must be preferred over isolated nanotubes for on-chip interconnect applications as the quantum resistance is large for an individual CNT. Arranging all individual CNTs in parallel to form a bundle structure reduces effective resistance. It can be noted that for long global wires, the interconnect delay is lower for carbon nanotube interconnects in comparison to Cu wires. In particular, an 80% reduction in delay is reported for CNT bundles at the global level. Enhancement in performance with advancements in technology node is achieved at a greater extent for global and intermediate interconnect levels than the local level. Thus, it is concluded that the CNT interconnect is more appropriate at advanced technology nodes.

Kulkarni and Khot (2010) had suggested different strategies adopted for maneuvering CNTs as interconnects. It is very well demonstrated that for giga scale devices and interconnect architectures the CNTs provide remarkable improvement in performance, but the key challenge lies in the implementation of tubes into interconnect technologies by judicially controlling the CNT position(s) and reproducing the tubes with acute precision at pre-specified locations. The various techniques discussed in this work include: 1) Non-directed synthesis approach and 2) directed synthesis approach. The non-directed approach involves post-growth positioning of tubes; thus, it is time lingering, not in harmony with existing silicon fabrication and procedures for manufacturing and hence highly inefficient. On the other hand, the directed synthesis approach is a simple one-step process involving a selectively controlled growth of tubes at predetermined locations on a substrate.

Naeemi and Meindl (2007) proposed a compact physics based circuit models of GNRs as a function of chiral vector, Fermi level, width and type of electron scattering at the edges. Comparative analysis of copper and CNT resistance are done with semiconducting and metallic GNRs for different Fermi energies at unity aspect ratios. Their investigation findings report that the metallic GNRs exhibit lower resistance in juxtapose to Cu and greater resistance than single-walled nanotubes for width dimensions less than 8 nm, but semiconducting GNR resistance becomes worse for the same dimensions of width. Higher Fermi energies lower the resistance because of more populated conduction channels and larger mean free paths. The merit of employing GNRs is that for a small thickness it makes 3–4 times less mutual capacitance between adjacent wires than copper. All of these factors result in reduced propagation delay, crosstalk noise, power dissipation and delay variation. Hence, it can be summarized that ultra-small thickness of GNRs can provide better performance than copper wires.

The investigations undertaken by Naeemi et al. (2004, 2005) in contradiction to Naeemi and Meindl (2005) show that a significant enhancement in interconnection speed can be achieved with ideal and densely packed CNT-bundle interconnects. The inferences drawn in the work of Naeemi et al. (2004,2005) are on the basis of optimistic assumptions that contacts in between metal and nanotubes are ideal in nature and nanotube bundles are completely metallic. However, the technological realities are totally different for carbon nanotube interconnects. In fact, the negative outcome of neglecting the imperfect metal-nanotube contact resistance resulted in significant error in value of resistance, leading to the wrong conclusion in Naeemi and Meindl (2005) that a flat array of CNTs is better suited than copper for local interconnects. The quantification of the critical issue of CNT bundle capacitance also raises serious concern. While Naeemi et al. (2005) ignore the calculation of capacitance by unjustifiably assuming that the capacitance for a CNT bundle is the same as that for copper interconnects, Naeemi and Meindl (2005) are unable to justify how the same interconnect analysis program can be used to extract capacitances for CNT bundles as for copper interconnects. The reduction in delay variation and overall digital circuit power dissipation by the use of a flat array of CNTs also does not reflect reality (Srivastava and Banerjee, 2005).

As reported the advancement in deep sub-micron technology resulted in a reduction of gate delays but the wire delays stay constant (Khursheed et al., 2018).

In an attempt to enrich the performance of a digital circuit, it is must to enhance the interconnect wire performance. For the copper interconnects, the trend shows a steep rise of resistivity with the scaling of dimensions. Futuristic on-chip interconnects, like GNR and CNT, are considered as a potential contender for copper replacement. Both CNT and GNR are based on their intrinsic nature and exhibit excellent electrical and mechanical properties with respect to copper.

Koo et al. (2007) discovered that for a local interconnect line, CNTs offer a smaller latency in signal compared to Cu. They have demonstrated a comparative analysis of optical and carbon interconnects with traditional Cu at all hierarchical semi-global and global wires. Their investigations highlighted the merits of novel technologies based on metrics such as latency, energy per bit/power dissipation and bandwidth density. A favorable characteristic of the SWCNT bundle is that it can be implemented in the same size scale as Cu wires (~50 nm); hence, can be used for all types of interconnects like local, semi-global and global applications. The supremacy of nanotubes is that it bears a higher value of conductivity due to its large mean free path and demonstrated higher EM tolerance and thermal conductivity than Cu (Chen and Friedman, 2006). Different graphs of CNT inductance along with components have been analyzed and the interpretation is made that shorter widths have a huge total inductance due to an increase in kinetic inductance. The results obtained show almost six times larger inductance to resistance ratio for CNT as compared to Cu, depicting a more pronounced impact of inductance in the case of the CNT bundle.

The research team led by Banerjee in Xu et al. (2008, 2009) analyzed the aptness of GNR to be used as futuristic interconnect material. To carry out an efficient analysis for the first time, edge specularity and intercalation doping effects are incorporated in both of the works. Besides this effect of width alteration, MFP for SLGNR is also explored. Further, to justify their analysis, a comparative analysis of GNRs with the conventional ones like Cu and CNTs is also done for the interconnect geometry. The research investigation summarizes that although GNRs are superior to CNTs from the fabrication point of view but fails in the performance point of view. The result shows the poor delay metrics and conductance metrics of GNRs as compared to its rival Cu and CNTs. At the 11 nm technology node, to make GNRs emerge as a genuine contender and not a delusive dream strategy of intercalation doping is applied on multilayer zigzag GNRs. This situation is witnessed only for intermediate and global interconnects, but at the local level, zigzag MLGNRs exhibit comparable performance to Cu and much better performance than tungsten, even with smooth edges assumption.

1.2.3 Review of Adopted Design Methodology for Interconnect Modeling

Zhao and Yin (2012) developed a circuit-oriented modeling techniques for both carbon nanotube (CNT) and graphene nanoribbon (GNR) interconnects. In lieu of the proposed model, the evaluation of transmission traits of single-walled CNT (SWCNT), multi-walled CNT (MWCNT) and multilayered GNR (MLGNR) interconnects along with their comparison with Cu is performed theoretically. To examine transmission characteristics and simultaneously demonstrate the effect of temperature distribution across an array of SWCNTs by accounting for SHE phenomenon so as to generate there transient thermal response by solving a 1D heat conduction equation. The suggested equivalent single conductor (ESC) circuit model for MLGNR shows an improvement in the case of a high-frequency application. It is also proclaimed as the beginning of the concept of "almost carbon integration" interconnects era for the development of new, future three-dimensional integrated circuits (3D ICs).

Previously, several experimental works (Murali et al., 2009; Naeemi and Meindl, 2009;Sarkar et al., 2011) involving graphene-based interconnects by and large concentrate only on the physics involved in quantifying the resistance of single-layer graphene. GNR interconnects are sculpted using a novel technique. Moreover, quite a lot of the existing literature available scrutinizes MLGNR interconnects by considering them as SC-MLGNRs only and completely overlooking the impact of top-contact layers and assuming all the layers connected in parallel. In Sarkar et al. (2011), analysis of the frequency response for both top- and side-contacts MLGNR wires is carried out. Computation of delay and energy on frequency response basis has been done. However, the models reported in Sarto and Tamburrano (2010) are purely empirical and it is trivial to extract interconnect performance factors and gain physical insights about the transient analysis of MLGNRs. The reported mode of Sui and Appenzeller (2009) comprises quantum resistance, quantum and electrostatic capacitance and kinetic and magnetic inductance along with a per unit length RLC parameters for these GNRs. The resistivity of GNRs has been obtained experimentally by authors in Adelman et al. (2014) by considering wider GNR interconnects range from 0.3 cm to 30 cm. A novel technique demonstrated for extracting resistivity of GNRs with narrow widths, the authors also highlight the effect of impurity scattering on resistivity of GNRs. Since graphene-based transistors can be fabricated by using narrow-width GNRs, therefore it can solve the problem of contact resistance. Authors have used transmission line analysis to find the frequency response of both SC-MLGNR and TC-MLGNR interconnects. They show the effect of quantum capacitance and kinetic inductance on the frequency response of MLGNRs. In Chai et al. (2008), TC-MLGNRs have been analyzed using a 2D resistor network to find its effective resistance for different interconnect widths. The authors extend their work and develop expressions for delay and energy delay product for MLGNRs interconnects.

Chuan Xu et al. calculated the conductance with the help of linear response landaur formula and tight binding model. They further laid their focus on the performance comparisons of GNRs with other interconnect materials. They extensively carried out a rigorous analysis on conductance of GNRs by studying the impact of various model parameters like edge specularity, mean free path, Fermi energy and bandgap. They concluded that for all widths in the case of global-level interconnects resistance of AsF5 doped zigzag multilayer GNR with smoother edges is comparatively less in value than copper, whereas a neutral zigzag multilayer GNR is found higher for the same case for lengths above 20 nm. They interpreted that both resistance and propagation delay of AsF_5 doped zigzag multilayer GNR with smoother edges is lower than copper and SWCNT for all lengths at 11 nm technology node for global length interconnects.

Atul K. Nishad et al. brought a turning point by proposing analytical time domain models for side-contact and top-contact multilayer graphene nanoribbon (MLGNR) interconnects. They have designed an optimum top-contact MLGNR interconnect that surpasses the performance with respect to Cu and optical interconnects. A closed-form analytical time domain model required to compute all performance factors including transfer function, delay and energy budget is developed. Closed-form expressions are proposed for the equivalent line resistance per unit length and transfer function for these interconnects. By various model parameters like interlayer conductance, Fermi energy and line geometry, they optimized the performance of top-contact MLGNR interconnects to match with that of side-contact MLGNR interconnects. The use of optimized top-contact MLGNRs gives better performance than copper interconnects and can eliminate the requirement of making side contacts (Nishad and Sharma, 2014). The authors have further explained that by raising the interlayer conductance value, TC-MLGNR delivers an equivalent performance as SC-MLGNRs. They have also introduced an improvement factor "g" that is the ratio of improved and normal interlayer conductance of TC-MLGNR interconnects. For superior performance of TC-MLGNR, a relatively lower "g" is required as it has also been shown that this factor "g" varies with variation in length. Finally, based on this model, a performance comparison is undertaken for the TC-MLGNR interconnects with Cu and optical interconnects. The TC-MLGNR interconnects match the theoretical performance of SC-MLGNR interconnects with a reduction in a complex fabrication process required for the side contacts.

Jiang-Peng Cui et al. extensively tackle the signal transmission analysis of MLGNR interconnects in their work (Cui et al., 2012) at the deep sub-micron technology nodes of 14 nm and 22 nm. In an attempt to examine the transient response, implementation of the equivalent single conductor (ESC) model is executed. The scrutiny Fermi level effects of MLGNR on the time delay of the rectangular pulse have also been perceived. Apart from this, both capacitive and inductive coupling among the neighboring GNR layers are taken into account. From the results of investigations obtained at the 22 nm technology node, it is concluded that the propagation time delay of the transmitted rectangular pulse diminishes with a decrement in Fermi level, whereas the transmission delay abruptly rises with an increment in length at the 14 nm technology node.

Kumar et al. (2012) have procured the physical models of effective resistance of m-GNR interconnects by taking into consideration the finite c-axis resistivity of MLGNR. It was recommended that for attaining excellent performance of MLGNR interconnect over SLGNR interconnect, the effect of screening has to be compensated. For achieving the said purpose adoption of doping approach is considered. To mitigate the delay and energy delay product in the case of MLGNR interconnects, the selection of the optimum number of layers are a must. It has been seen that the number of graphene layers are proportional to the length, width and edge-scattering probability. A comparison of MLGNR and Cu interconnects at the 9.5 nm node shows that MLGNR performed better in terms of delay if the GNR edges are smooth and with the combination of minimum-sized drivers (Kaushik et al., 2007).

Nasiri et al. (2010) investigated a TLM to carry out a time domain analysis of GNR-based interconnect structures. The strategy adopted by them for analyzing the dependence of (MLGNR) interconnects on their dimensions and contact resistances is by applying the transmission line model and linear parametric expression for calculating the transfer function of a driver–MLGNR interconnect–load (DIL) system. For the execution of the aforementioned strategy, a sixth-order linear parametric approximate relation is applied to simulate the step response. The results of Nyquist stability analysis points out the instability and delay rise with an increase in lengths, widths or contact resistances. The time domain output responses and Nyquist plots interpret reduced propagation delays for MLGNRs than SWCNT bundle interconnects of congruent dimensions.

1.2.4 Need for Repeater Insertion

Bakoglu and Meindl (1985) judicially proposed a model for an interconnect structure time delay that incorporates all the impacts caused by scaling of transistor dimensions and interconnect dimensions to shrink the on-chip feature size. A comparative study on the delays of polysilicon lines, aluminium and WSi_2 was executed. It is observed that with an increase of the chip dimensions, there will be a reduction in minimum feature size, as a consequence of which impact of R_{int} becomes prominent and delay shoots abruptly. Hence, the major challenge is to lower the effect of both capacitances as well as resistances in order to mitigate the delay. The dual technique adopted for mitigating the propagation delay is to properly suppress the interconnecting wire resistance by using only aluminum wires for long-distance communication and by forming a multilayer of interconnects with thicker and wider lines in the upper levels. The second technique undertaken is to enrich the performance of a driver circuit by either using cascaded drivers that increase in size sequentially until the last device was large enough to drive the line. In addition to that technique of repeaters, insertion by dividing the interconnections into smaller subsections is followed.

El-Moursy and Friedman propounded that an optimum-tapered structure can be incorporated for interconnect structures to effectively suppress power loss and energy consumption. For global interconnect structures that are operating at a frequency much greater than GHz, the inductive impact cannot be overlooked. Wire tapering is a most suitable approach adopted that can effectively mitigate signal propagation delay in the RLC modeled on chip circuits. Transition time at the load gate is reduced by a tapered interconnect wire. Capacitance of the line also decreases, thereby reducing both short-circuit power of the load gate and dynamic power of the driver. Thus, overall total power dissipation of the tapered interconnect is less than the total power dissipation of a uniformly sized interconnect wire.

Giustiniani and Tucci (2010) presented a thorough analysis on the performance of a semiglobal interconnect structure. The analysis is undertaken on the basis of a densely packed single-walled carbon nanotube (SWCNT) bundle. The consequences of impact of non-linear distributed parameters that are found in TLM are also analyzed. It was investigated that the high-frequency behavior of interconnects is largely impacted by random distribution of metallic nanotubes within the CNT bundle, which alters the capacitive effect. The interpolation of equispaced repeater buffers is determined by kinetic inductance and causes a reduction in time delay. The impact of the presence of large contact resistance in between the CNT bundle and lumped circuit causes a decrease in bandwidth. This is a major hindrance for the evolution of CNT-based interconnect structures for high-speed applications.

Adler and Friedman (1998) proposed that signal propagation delay is hindering the performance of large integrated circuits due to an increase in resistance of long interconnect wires. The strategy of repeater insertion decreases the effect of quadratic rise in delay. It lowers the power by reducing the short circuit current. The short circuit power dissipation for a repeater inserted circuit is calculated and an equation is presented in the work. Discussion on dynamic power is also done. It is observed that a repeater inserted to any RC line can significantly decrease the signal propagation delay.

Chandel et al. (2005) presented a systematic bibliographic analysis of repeater-loaded global interconnect structure for very large-scale integrated circuits. These structures are analyzed on the basis of delay and power design metrics. A voltage-scaled repeater has an added advantage in designing ultra-low-power VLSI circuits. They are instrumental in reducing the chip area as well as achieving a proper trade-off between delay and power. Several techniques adopted for inserting optimum uniform-sized buffers in between very long global wires is reviewed in this work.

Banerjee and Mehrotra investigated that despite the fact that a repeater insertion optimizes delay but also adds to power dissipation. A method is presented in this work to find the buffer size and minimum interconnect length that can optimize the total power dissipation due to buffer insertion for a given acceptable delay. Short circuit and leakage power must be taken into account for power calculation. Device scaling results in an exponential increase of leakage power and decrease of switching power.

Alpert et al. (1999) demonstrated that repeaters can also be used to reduce the crosstalk noise with an increase of power dissipation. PDP is a better criterion to quantify the optimum number of repeaters so as to reduce the delay, power and crosstalk. Repeater insertion can nullify the negative effects of wire sizing on delay. Optimization of repeaters for delay minimization for scaled voltage leads to area and power saving. Temperature analysis of two different repeaters for global copper interconnects was also done.

Various optimizing techniques adopted to date by researchers for VLSI interconnects can help in a better way to reduce delay, power dissipation and crosstalk, yet a high-speed and power-efficient repeater can further reduce the number of repeaters.

1.2.5 Performance Tuning of On-Chip Interconnects on the Basis of Design Metrics

Rakheja et al. (2013) in their procured manifesto on the performance of graphene nanoribbons (GNR) as VLSI interconnects, examined the physical models of electron MFP, diffusion coefficient, resistance p.u.l. and analyzes the mobility for bulk and narrow GNRs as a function of wire size, edge roughness and Fermi energy shift. Considering theoretical values of spin-orbit coupling (SOC) caused due to ripples in graphene, they have obtained spin-relaxation length in graphene. The models procured in their work are set as a benchmark for GNR interconnects with respect to classical Cu/low-k interconnects in both electrical and spintronics domains. An analogy between state-of-the-art MLGNR interconnects and Cu at the 9.5 nm technology node reveals that delay latency in MLGNR is less than Cu only if the GNR edges are smooth and interconnects are driven by minimum-sized drivers. The presence of adatoms in graphene restricts the spin-relaxation lengths, making it unsuitable for interconnect applications. Further, MFP of electrons limits the speed of spin interconnects. The speed deteriorates with an increase in edge roughness and diminution in interconnect width. There is inconsistency in the energy saving of electrical and spin interconnect with the change in interconnect length. Thus, at the local and intermediate level, spin interconnects are preferred, while at the global level, electrical ones will only rule. The primary reason behind this interpretation is the hike in energy consumption of spin interconnects with an increase in length. For future generation devices, it is suggested to use a hybrid of both spin and electrical interconnects (Iraei, 2017).

Debaprasad Das and Hafizur Rahaman investigate the reliability of gate oxide in GNR interconnects and also quantifies the impact of crosstalk noise for graphene nanoribbon interconnects at the 16 nm technology node. As per the investigation findings of this work, they concluded that near-end crosstalk is found to be greater in GNR compared to Cu and MWCNT interconnects, but the far-end crosstalk noise is less in GNR compared to Cu and MWCNT. Moreover, GNR has a less gate oxide failure rate than Cu.

Pandya et al. (2012) in their research presented a precise hierarchical model of m-CNT bundle interconnects. Depending on the hierarchical style chosen, varied structures are procured for distinct arrangements of SWCNT and MWCNT. By concatenating ESC models of SWCNT and MWCNT, they developed an equivalent model of mixed CNT bundles. Later on, based on the proposed structures of CNT bundles, a comparative analysis of delay under the influence of dynamic crosstalk is executed. The results obtained justify the improvement in crosstalk-induced propagation delay of structures in which single-wall carbon nanotubes are at the center and multi-wall carbon nanotubes are at the periphery.

Narsimha Reddy et al. in this research work have performed a simulation on MLGNR interconnects to calculate the crosstalk delay. For the simulation purpose, they have considered a two coupled wire bus architecture integrated with CMOS drivers and analyze the in-phase and out-of-phase delays (Reddy et al. 2012). Results show that if two wires are in-phase, then the coupling capacitance has no effect on delay, but the worst case is observed when signals are out-of-phase on each line. Apart from this impact, the Miller Coupling factor on the coupling capacitance between aggressor and victim lines is also investigated. Lastly, it was concluded that in-phase and out-of-phase crosstalk delays show improvement by 4.75% and 18.04%, respectively, for MLGNR with a greater number of layers in comparison to the fewer layers. The reason behind this is an increment in conducting channels with an increment in layers and width of MLGNR.

Cui et al. (2012), on the basis of the transmission line model, investigate the crosstalk-induced signal transmission for multilayered graphene nanoribbon interconnects. Their team from the investigation obtained the output response of the DIL system using the transfer function. The analysis of the effect of the Fermi level on the signal transmission was also carried out.

Zhao and Yin (2014), along with his team, performed a comparative analysis on different types of interconnects, namely MLGNR, SWCNT, MWCNT and Cu. From their analysis, they concluded that even for the case of maximum crosstalk, the MLGNR interconnects show the superiority over other wire materials used for interconnect modeling. In their research work, Sahoo and Rahaman (2014) investigated the influence of variation in line resistance of MLGNR on the induced performance parameters due to crosstalk. The simulation analysis was done on 11 nm and 8 nm technology nodes of long global-level wires as well as on medium-level interconnect wires. The authors had taken the mfp independent of width by considering perfectly smooth edges of multilayered nanoribbons of graphene. It has been seen that no matter what technology node is taken, in all scenarios, the MLGNR interconnects demonstrate superior performance than Cu wires.

1.3 SUMMARY

This chapter presents an overview of interconnect device physics and related issues. It begins with an historical review on high-speed interconnects wire modeling. To get a deep insight into the versatility of the interconnect research area, readers are incentivized to go through the variety of literature available on aforementioned topics. In a nutshell, this chapter lays the thoroughfare for more detailed analysis to be followed in the ensuing chapters.

1.4 MOTIVATION

As predicted by ITRS, with the passage of time, the field of interconnects has emerged as a major research area for both industry and academics. New-age VLSI chips contain billions of transistors embedded onto it. With the advancement in nanometer technology, the miniaturization of IC takes place resulting in more and more functional blocks assimilated on a single chip. These functional blocks and the devices on them are all connected to each other with the help of interconnect vias and wires. A bottleneck situation is being created as an outcome of dense interconnection due to high chip complexity. This trivial condition makes the field of interconnects a tantalizing research topic.

This book will be a trailblazer for the researchers and graduate students specifically working in the area of low-power VLSI circuits. It mainly focuses on resolving the issues that arise due to scaling of on-chip interconnect dimensions. It also sheds light on the application of carbon nanomaterials for modeling VLSI interconnect and buffer circuits; besides describing the electrical, thermal and mechanical properties, and structural behavior of these materials.

1.5 BOOK OUTLINE

This book ferrets out the issues encountered by VLSI circuits due to scaling of on-chip interconnect dimensions. The primary goal of this manuscript is to bring forth a comprehensive analysis of ultra-low-power high-speed nano-interconnects based on different facets like material modeling, circuit modeling and the adoption of repeater insertion strategies and measurement techniques is pertinent to current research scenario.

The book is comprised of four chapters.

Chapter 1 This chapter gives an overview of the difficulties encountered in terms of interconnect performance deterioration with the advancement in VLSI industries and technology scaling. It also presents an extensive literature review on the background of interconnect evolution, their growing trend with upcoming future technologies and different strategies adopted by various researchers to handle the existing problem. Just a glance at this introductory chapter will be sufficient for well-informed readers.

Chapter 2 This chapter gives an introduction on types of interconnects, highlighting their hierarchical structure in ASIC. It emphasizes the use of graphene (CNT and GNR) based interconnect as a substitute to conventional interconnect materials. In the end, various existing models of interconnect structures are presented and a description about their electrical behavior is also given.

Chapter 3 Repeater buffer insertion is a trusted approach for reducing the delay and enhancing the performance of on-chip interconnects, but their exponential increase in number for very long wire lines is becoming a hurdle in producing energy-efficient ICs. This chapter initially provides a background of traditional repeater interpolation methodology implemented in the interconnect network to mitigate propagation delay. It then mentions the various styles of buffer insertion to overcome the limitation of area overhead and power consumption. It further suggests the use of state-of-the-art device modeling to realize high-speed ultra-low-power smart buffers. The structure of novel CNTFET/GNRFET-based buffers and their compatibility with the existing as well as emerging wire technology and device technology is also explained in the end.

Chapter 4 This final chapter concludes with its first section presenting the brief discussion on signal integrity challenges encountered when modeling state-of-the-art nano interconnects. As feature size decreases due to scaling of device dimensions, the reliability is compromised. The decrease in reliability is due to electromigration-induced problems. At a very high frequency of operation, the tightly packed nano interconnects produce transient crosstalk. Furthermore, analysis on how the crosstalk noise strongly influences the signal propagation delay and causes the circuit malfunction or functional failure is done. As the crosstalk noise causes signal overshoot, undershoot and ringing effects, hence it is enviable to develop a perfect pedagogical model for analytically analyzing the crosstalk effects occurring in nano interconnects.

For this reason, the latter part of this chapter presents a precise and time-efficient model of FET-gate-driven coupled (copper/CNT/GNR) nano interconnects in order to commensurate the crosstalk-incited performance analysis of nano interconnects. The model discussed is formulated by finite difference time domain (FDTD) methodology for the DIL system by assimilating boundary conditions for all the three types of interconnect materials. The FET driver is modeled by either nth power law model by considering the finite drain conductance parameter.

REFERENCES

Adelman, Christoph, Liang Wen, Antony Peter, Yong Siew, Kristof Croes, J. Swerts, M. Popovici, et al., 2014. "Alternative Metals for Advanced Interconnects," 2014 IEEE International Interconnect Technology Conference/Advanced Metallization Conference, IITC/AMC 2014, 173–176. doi: 10.1109/82.673643.

Adler, Victor and E. G. Friedman, "Repeater Design to Reduce Delay and Power in Resistive Interconnect." *IEEE Transactions on Circuits and Systems II: Analog and Digital Signal Processing* 45, no. 5 (1998): 607–616. doi: 10.1109/82.673643.

Alpert, J. Charles, Aniruth Devgan, and Stephen T. Quay. "Buffer Insertion for Noise and Delay Optimization." *IEEE Transactions on Computer-Aided Design of Integrated Circuits and Systems*, 18, no. 2 (1999): 1633.

Bakoglu, H. B. and J. D. Meindl. "Optimal Interconnection Circuits for VLSI." *IEEE Transactions on Electron Devices,* ED 32, no. 5 (1985): 903–909.

Banerjee, Kaustav and Amit Mehrotra. "A Power Optimal Repeater Insertion Methodology for Global Interconnects in Nanometer Designs." *IEEE Transactions on Electron Devices*, 49, no. 11.

Banerjee, Kaustav and Navin Srivastava. "Are Carbon Nanotubes the Future of VLSI Interconnections?." in Proceedings of 43rd Annual Design Automation Conference, San Francisco, 809–814, 2006.

Bernd, Hoefflinger. "ITRS: The International Technology Roadmap for Semiconductors. Chips 2020: A Guide to the Future of Nanoelectronics." *The Frontiers Collection* 161 (2012). doi: 10.1007/978-3-642-23096-7_7.

Bohr, Mark T. "Interconnect Scaling—The Real Limiter to High Performance ULSI." in Proceedings of International Electron Devices Meeting, Washington, DC, 1995, 241–244.

Ceyhan, Ahmet and Azad Naeemi. "Cu Interconnect Limitations and Opportunities for SWNT Interconnects at the End of the Roadmap." *IEEE Transactions on Electron Devices* 60 (2013): 374–382.

Chai, Yang, Philip C. H. Chan, Yunyi Fu, Y. C. Chuang, and C. Y. Liu. "Copper/Carbon Nanotube Composite Interconnect for Enhanced Electromigration Resistance." in Proceedings of 58th Electronic Components and Technology Conference, 2008 : ECTC 2008, Disney's Contemporary Resort, Lake Buena Vista, Florida, USA, 27–30 May 2008, 412.

Chandel, Rajeevan, S. Sarkar, and R. P. Agarwal. "Repeater Insertion in Global Interconnects in VLSI Circuits." *Microelectronics International* 22, no. 1 (2005): 43–50. doi: 10.1108/13565360510575549.

Chen, Guoqing and E. G. Friedman. "Low-power Repeater Driving RC and RLC Interconnects with Delay and Bandwidth Constraints."*IEEE Transactions on VLSI Systems*14, no. 2 (2006): 161–172.

Cui, Jiang-Peng, Wen-Sheng Zhao, Wen-Yan Yin, and Jun Hu. "Signal Transmission Analysis of Multilayer Graphene Nano-ribbon (MLGNR) Interconnects." *IEEE Transactions on Electromagnetic Compatibility* 54, no. 1 (2012): 126–132.

Das, Debaprasad and Hafizur Rahaman. "Crosstalk and Gate Oxide Reliability Analysis in Graphene Nanoribbon Interconnects." in Proceedings of International Symposium on Electronic System Design, Kochi, Kerala, India, 182–187, 19–21 December 2011.

El-Mousry, Magdy A. and Eby G. Friedman, "Optimum Wire Tapering for Minimum Power Dissipation in RLC Interconnects", IEEE International Symposium on Circuits and Systems (2006): pp. 4. doi: 10.1109/ISCAS.2006.1692628.

Giustiniani, A. and V. Tucci. "Modeling Issues and Performance Analysis of High-Speed Interconnects Based on a Bundle of SWCNT." *IEEE Transactions on Electron Devices* 57 (2010): 1978–1986.

Ho, Ron, Kenneth W. Mai, and Mark A. Horowitz. "The Future of Wires." *Proceedings of the IEEE* 89, no. 4 (2001): 490–504.

International Technology Roadmap for Semiconductors (ITRS) Reports, 2013 [online]. Available: http://www.itrs.net/reports.html.

Iraei, Rouhollah Mousavi. "Electrical-Spin Transduction for CMOS-Spintronic Interface and Long-Range Interconnects." *IEEE Journal on Exploratory Solid-State Computational Devices and Circuits* (2017). doi: 10.1109/JXCDC.2017.2706671.

Kapur, Pawan, James P. McVittie, and Krishna C. Saraswat. "Technology and Reliability Constrainted Future Copper Interconnects-Part I: Resistance Modeling." *IEEE Transactions on Electron Devices*49, no. 4 (2002): 590–597.

Kaushik, Brajesh Kumar, Saurabh Goel, and Gaurav Rauthan. "Future VLSI Interconnects: Optical Fiber or Carbon Nanotube a Review," *Microelectronics International*, 24, no. 2 (2007): 53–63. doi: 10.1108/13565360710745601.

Khursheed, Afreen, Kavita Khare, and Fozia Z. Haque. "Performance Tuning of VLSI Interconnects Integrated with Ultra Low Power High Speed DSM Repeaters." *Journal of Nano and Optoelectronics* 13, no. 12 (2018): 1797–1806. doi: 10.1166/jno.2018.2507.

Khursheed, Afreen, Kavita Khare, and Fozia Haque. "Designing of Ultra-low-power High-speed Repeaters for Performance Optimization of VLSI Interconnects at 32 nm." *International Journal of Numerical Modelling: Electronic Networks, Devices and Fields* 32 (2018). doi: 10.1002/jnm.2516.

Koo, Kyung-Hoae, Hoyeol Cho, Pawan Kapur, and Khrishna C. Saraswat. "Performance Comparisons Between Carbon Nanotubes, Optical, and Cu for Future High-performance

On-chip Interconnect Applications." *IEEE Transactions on Electron Devices* 54, no. 12 (2007): 3206–3215.

Kulkarni, Chinmay and Ajinkya Khot. "Carbon Nanotubes as Interconnects." *Indian Journal of Pure & Applied Physics* 48 (2010): 305–310.

Kumar, Vachan, Shaloo Rakheja, and Azad Naeemi. "Performance and Energy per Bit Modeling of Multilayer Graphene Nanoribbon Conductors." *IEEE Transactions on Electron Devices* 59, no. 10 (2012): 2753–2761.

Li, Hong, Wen-Yan Yin, and Jun-Fa Mao, "Modeling of Carbon Nanotube Interconnects and Comparative Analysis with Cu Interconnects," Proceedings of Asia-Pacific Microwave Conference, Yokohama, 1361–1364, 12–15 December 2006. doi: 10.1109/APMC.2006.4429659.

Hong Li, Chuan Xu, Navin Srivastava, and Kaustav Banerjee et al. ''Carbon Nanomaterials for Next-Generation Interconnects and Passives: Physics, Status, and Prospects,'' *IEEE Transactions on Electron Devices* 56, 9, (2009): 1799–1821. doi: 10.1109/TED.2009.2026524.

Murali, Raghunath, Yinxiao Yang, Kevin Brenner, Thomas Beck, and James D. Meindl, "Breakdown Current Density of Graphene Nanoribbons," *Applied Physics Letters*, 94, no. 24 (2009): 243114–243114-3. doi: 10.1063/1.3147183.

Murugeswari, P., A. P. Kabilan, S. Rohini, and P. Pavithral, "Analysis of Carbon Nano Structures for On-chip Interconnect application" *ARPN Journal of Engineering and Applied Sciences*, 10 (2015): 2702–2706.

Naeemi, Azad and James D. Meindl, "Monolayer Metallic Nanotube Interconnects: Promising Candidates for Short Local Interconnects," *IEEE Electron Device Letters*, 26, no. 8 (2005): 544–546

Naeemi, Azad and James D. Meindl, "Conductance Modeling for Graphene Nanoribbon (GNR) Interconnects," *IEEE Electron Device Letters*, 28, no. 5, 428–431, 2007.

Naeemi, Azad and James D. Meindl, "Compact Physics-based Circuit Models for Graphene Nanoribbon Interconnects," *IEEE Transactions on Electron Devices*, 56 (2009).

Naeemi, Azad, Reza Sarvari, and James D. Meindl, "Performance Comparison Between Carbon Nanotube and Copper Interconnect for Gigascale Integration (GSI)," *IEEE Electron Device Letters*, 26, no. 2 (2005): 84–86,

Naeemi, Azad, Reza Sarvari, and James D. Meindl, "Performance Comparison Between Carbon Nanotube and Copper Interconnects for GSI", *IEEE International Electron Devices Meeting*, 699–702, 13–15 December 2004.

Nasiri, Saeed Haji, Mohammad Kazem Moravvej-Farshi, Rahim Faez. "Stability Analysis in Graphene Nanoribbon Interconnects." *IEEE Electron Device Letters* 31, no. 12 (2010): 1458–1460.

Nishad, Atul K. and Rohit Sharma, "Analytical Time-domain Models for Performance Optimization of Multilayer GNR Interconnects," *IEEE* 20, 1, 17–24, 2014.

Pandya, Nisarg D., Manoj Kumar Majumder, Brajesh Kumar Kaushik, and Sanjeev Kumar Manhas, "Dynamic Crosstalk Effect in Mixed CNT Bundle Interconnects," *Electronics Letters* 48, no. 7 (2012): 384–385.

Rabaey, Jan M., Anantha P. Chandrakasan, and Borivoje Nikolic, "Digital Integrated Circuits", 2nd edition, ISBN-13: 978-0130909961, 2003.

Rakheja, Shaloo, Vachan Kumar, and Azad Naeemi. "Evaluation of the Potential Performance of Graphene Nanoribbons as On-Chip Interconnects." *Proceedings of the IEEE* 101, no. 7 (2013): 1740–1765.

Reddy, Narasima K., Manoj Kumar Mazumdar, Brajesh Kumar Kaushik, Sanjeev Kumar Manhas, and Bulusu Anand. "Dynamic Crosstalk Effect in Multi-layer Graphene Nanoribbon Interconnects." in Proceedings of IEEE International Conference on Communications, Devices and Intelligent Systems, Kolkata, 472–475, 28–29 December 2012.

Sahoo, Manodipan and Hafizur Rahaman, "Impact of Line Resistance Variations on Crosstalk Delay and Noise in Multilayer Graphene Nano Ribbon Interconnects". In: Proceedings of the International Symposium on Electronic System Design (ISED) 2014): 94–98.

Sarkar, Deblina, Chuan Xu, Hong Li, and Kaustav Banerjee. "High-frequency Behavior of Graphene Based Interconnects Part I: Impedance Modeling." *IEEE Transactions onElectron Devices*, 58, no. 3 (2011): 843–852.

Sarto, M. Sabrina and Alessio Tamburrano. "Comparative Analysis of TL Models for Multilayer Graphene Nanoribbon and Multiwall Carbon Nanotube Interconnects." in Proceedings of 2010 IEEE International Symposium on Electromagnetic Compatibility (EMC), pp. 212–217, 2010.

Srivastava, Navin and Kaustav Banerjee, "Performance Analysis of Carbon Nanotube Interconnects for VLSI Applications." in Proceedings of IEEE/ACM International Conference on Computer Aided Design, 383–390, 6–10 November 2005.

Srivastava, Navin, Rajiv V. Joshi, and Kaustav Banerjee. "Carbon Nanotube Interconnects: Implications for Performance, Power Dissipation and Thermal Management." *Electron Devices Meeting, IEDM Technical Digest. IEEE International* 252 (2005): 249.

Steinhogl, W., G. Schindler, G. Steinlesberger, M. Traving, and M. Engelhardt "Comprehensive Study of the Resistivity of Copper Wires with Lateral Dimensions of 100 nm and Smaller." *Journal of Applied Physics* 97, no. 2 (2005): 023706-1–023706-7.

Sui, Yang and Joerg Appenzeller. "Screening and Interlayer Coupling in Multilayer Graphene Field-effect Transistors." *Nano Letters* 9, no. 8 (2009): 2973–2977.

Weste, Neil and David Harris "CMOS VLSI Design, A Circuits and Systems Perspective". 3rd Edition, Pearson Education, 2008.

Xu, Chuan, Hong Li, and Kaustav Banerjee. "Graphene Nano-ribbon (GNR) Interconnects: A Genuine Contender or a Delusive Dream?." in Proceedings of IEEE International Electron Devices Meeting, San Francisco, 1–4, 15–17 December 2008.

Xu, Chuan, Hong Li, and Kaustav Banerjee, "Modeling, Analysis, and Design of Graphene Nano-ribbon Interconnects." *IEEE Transactions on Electron Devices* 56, no. 8 (2009): 1567–1578.

Zhao, Wen Sheng and Wen Yan Yin. "Carbon-Based Interconnects for RF Nanoelectronics." *Wiley Encyclopaedia of Electrical and Electronics Engineering*, Piscataway, NJ: IEEE – Institute of Electrical Electronics Engineers Inc., 2012.

Zhao, Wen-Sheng and Wen-Yan Yin. "Comparative Study on Multilayer Graphene Nanoribbon (MLGNR) interconnects." *Transactions on Electromagnetic Compatibility* 56, no. 3 (2014): 638–645.

2 Interconnect Modeling

2.1 INTRODUCTION

Rapid development in the field of VLSI fabrication pushes the chip industry to enter the era where minimum feature size of leading processes is much below 1 micrometer. Such processes are termed deep submicron (DSM) processes. In state-of-the-art DSM technology designs with further advancement in process technology, the signal propagation delay within the interconnect network becomes a constraint over the transistor delay. Interconnect networks are clusters of wires providing electrical connections at all metallic levels between devices embedded within the chip, as shown in Figure 2.1. Thus the focus of ASIC design flow is gradually shifting from logic optimization towards interconnect optimization.

As far as gate scaling is concerned, improvement in device performance is obtained with a reduction in length of gate channel, thickness of gate dielectric and depth of junction. Contradictory to this, scaled interconnect wires suffer from a hike in resistance due to reduction in cross-section area of the chip; moreover, if conductor height is not proportionately reduced with spacing so as to maintain a proper aspect ratio then capacitance also increases. To lessen abnormally large delays, long interconnect dimensions are scale down more sluggishly than the device dimensions.

2.2 TYPES OF INTERCONNECTS ON A CHIP

For digital circuit designs, the effect of cramming large number of components on a single chip results in device performance to overtake interconnect capabilities. The problem is quite appalling for memory cells with the most aggressive metal pitch, and highest aspect ratio among all semiconductor devices. Interconnects are stacked with dielectric material between two layers or between one layer to the transistor, as shown in Figure 2.1. The interconnect wire structure is alienated by interlayer dielectrics (ILDs) from one metal level to another metal level and sequestered by inter-metal dielectrics (IMDs) within the same metal level (Elgamel and Bayoumi, 2003).

The inceptive fabrication processes have a solitary metal layer, but with remarkable improvement in chemical mechanical polishing techniques it became far more realistic to manufacture IC on a smaller footprint with multi-metal layers. Taking into consideration the example of Intel metal stacks, a 65 nm process generally has 8–10 metal layers and it has been observed that the layer count is increasing at a rate of about one layer every process generation. It can be revealed from the Figure 2.2, showing a cross section of Intel's metal stack that typically a 90 nm process has six metal layers, whereas a 45 nm process has eight metal layers on the top of transistors.

DOI: 10.1201/9781003104193-2

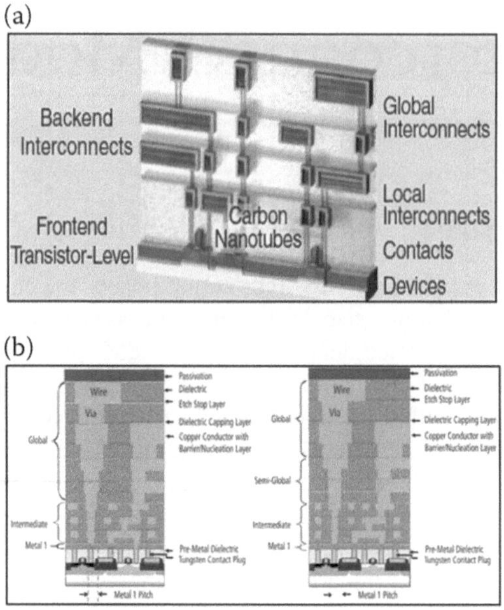

FIGURE 2.1 (a) Cross-sectional view of an IC showing CNT-based interconnects and devices; (b) illustration of cross sections of hierarchical scaling depicting different interconnect levels in MPU (left) and ASIC (right) (Permission, 2017).

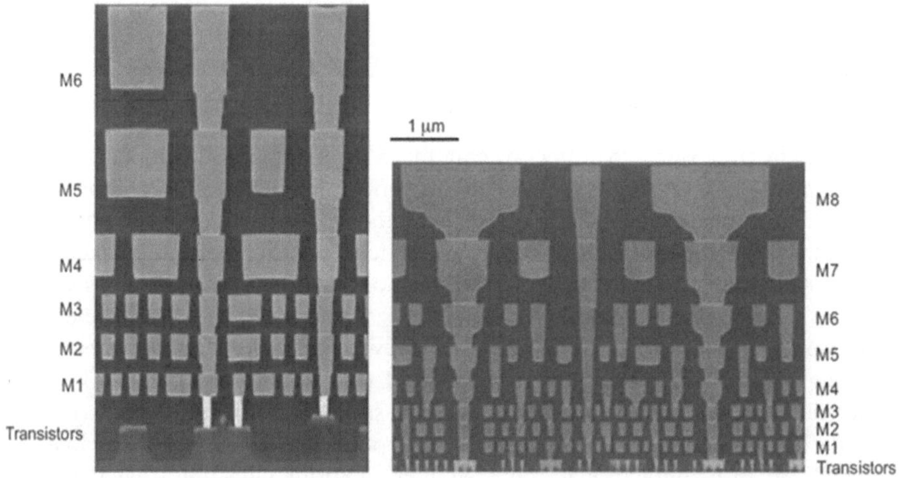

FIGURE 2.2 Cross-section view of Intel's metal stack for 90 nm and 45 nm process.

The upper metal layer is utilized to route data signal, clock, power supply and ground throughout the chip. Lower layers are employed to provide connection between gates of transistors. Hence, it can be concluded that the top-most layers must be designed for handling the greater currents required for increased power

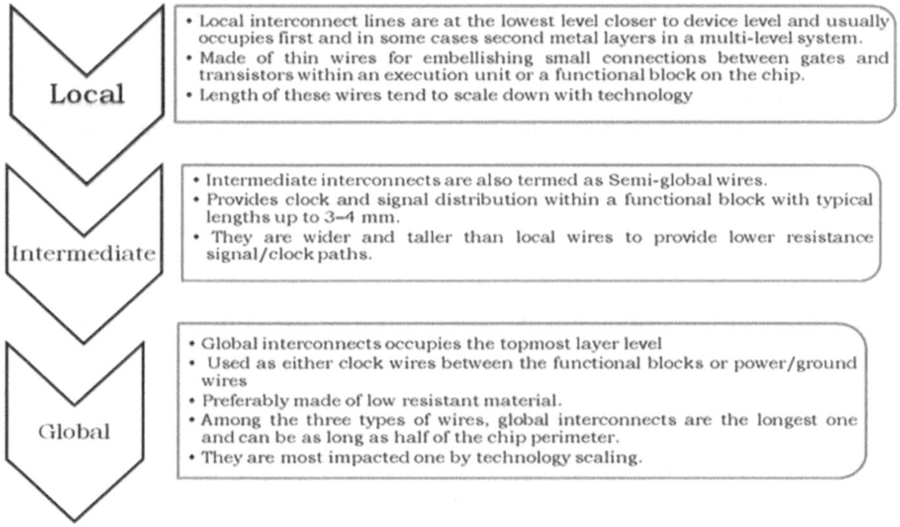

FIGURE 2.3 Types of interconnects.

consumption at a reduced voltage supply, whereas the lowest layers must be designed for reduced pitch and thickness; therefore, the top-most layers are thicker and fabricated using low-resistant material.

On the basis of length and cross-sectional dimensions, interconnecting wires are classified as local, intermediate and global as illustrated by the chart in Figure 2.3. A brief description of these three different interconnect layers are as follows:

a. Local: The local interconnects are wires that are very thin and primarily employed for providing an interface in between logic gates and transistors within an arithmetic logic execution unit or a functional block on the ASIC. Local interconnects extend up to a few spans and lie in first and second metal layers in a multi-level system. With advancement in process technology and further minimization of device features, the length of interconnect wires also scales down.
b. Semi-global/Intermediate: In terms of span of length, the semi-global or intermediate interconnects lie in between local and global wire length. The typical length is 3 to 4 mm. As compared to local interconnect rails, the intermediate rails are thicker wires and also taller in height to maintain the aspect ratio. Semi-global wires are employed for dispensing a timing clock and data signal distribution within a functional block on the ASIC.
c. Global: Global wires are employed for dispensing a timing clock and data signal distribution in between a functional block on the ASIC, whereas semi-global provides within the functional block on the ASIC. In addition to this, it is also used to provide power and ground to all devices and functional blocks onto an ASIC. The typical length of a global interconnect lies in a range greater than 4 mm and in some cases found to be as greater than half of the chip perimeter.

2.3 INTERCONNECT MODELS: CONSTRAINTS AND REQUIREMENTS

As the feature size continues to shrink while the clock frequency continues to increase, this opens the door for technology to step into a new era thereby yielding faster and denser chips with ever-increasing functionality. Previously, the IC circuit designers considered only device models in circuit simulations, whereas interconnect modeling were neglected. To further increase the functionality and subsequent number of transistors, a larger number of interconnect levels are required that lead to multi-layer interconnect systems nowadays. Henceforth, an accurate and efficient modeling of interconnecting lines is one of the critical aspects of both verification and the circuit design process engineering. Modeling of on-chip interconnects primarily focuses on the generation of 2D/3D interconnect model libraries using the geometry and material characteristics from foundry's electrical design rules. Quantifying the impact of interconnecting wires on the circuit behavior is a must for the combined optimization of circuits and interconnects.

To study the electrical behavior of these interconnecting wires, various types of widely proposed models for single-ended signal transportation are shown in Figure 2.4. The criteria of selecting a suitable model necessary for accurate and efficient circuit analysis is briefly described in this section. These models are bifurcated as lumped element models and distributed models depending on length of wire in regards to signal wavelength, operating frequency and condition of signal.

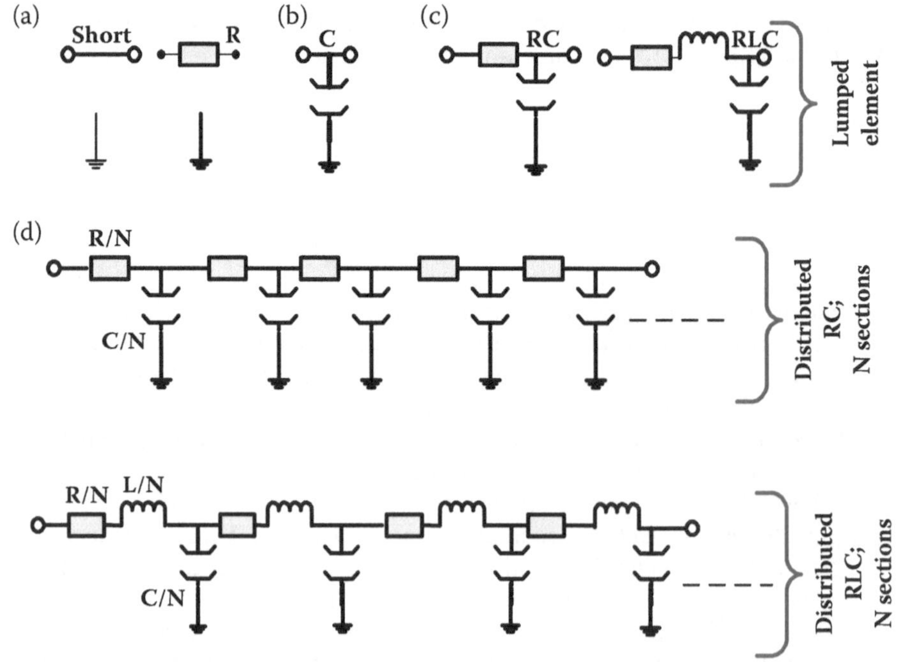

FIGURE 2.4 Interconnect models.

In the formative years of chip design, the interconnecting wires had insignificant values of parasitic impedances compared to gate impedances, due to large gate size (length and width). Thus, initially the wires were taken as a short circuit [shown in Figure 2.4(a)]. However, with the appearance of the micrometer technology era, the wire capacitances become commensurable to load capacitance and henceforth modeled as a single lumped capacitance [shown in Figure 2.4(b)]. In this case, the wire delay is computed as driver resistance times the total load capacitance. But for very long global interconnecting wires, having reduced cross sectional area, due to technology scaling effect, the lumped capacitance model becomes inaccurate and generates erroneous results. Hence, the need arises for more elaborate models which also reckon the effect of line resistance. Now the interconnecting wires are represented as a lumped RC network. Two commonly used circuits to represent RC networks are π and T circuits [depicted in Figure 2.4(c)]. Here, both π and T-type RC networks can model interconnecting wires with less than a 3% delay error. It has been found that these RC models hold good only for low to medium frequency of operation, but as the operating frequencies become higlayers on insulaô€€Ÿng substratesh (GHz range), they become inadequate for precise characterization of waveform properties along a wide interconnecting wire. The use of high frequency increases the value of $|j\omega L|$, making the effect of parasitic line inductance noticeable in the calculation of delay.

Moreover, with the advent of the nanometer era and introduction of novel materials, the electrical length of interconnecting lines becomes a significant fraction of the operating wavelength, resulting in unwanted signal distortion, particularly at very high operating frequencies. In such scenarios, the conventional interconnect lumped RLC model fails. A trade-off for accuracy and efficiency is done in selecting the suitable model. Hence, the distributed RLC model, which is proven to be most accurate [shown in Figure 2.4(d)], is opted for expressing the electrical behavior of wire. This distributed transmission line analogy-based RLC model assumes signal propagation to be a wave propagating over a long, wide wire and can be analyzed with the help of Maxwell's equation.

2.4 INTERCONNECT MATERIAL CHALLENGES AND TACTICAL ELUCIDATION

As predicted by the Semiconductor Industry Association's Technology Road map, downscaling of integrated circuits in accordance with Moore's Law will result in several performance issues [International Technology Roadmap for Semiconductors (ITRS), 2013]. At the lower technology node, the interconnect delays become more prominent than gate delays. As discussed in a previous section, interconnecting wires do not act as simple resistors at higher frequencies but also have parasitic capacitance and inductance associated with them. These impedance parameters are material-dependent and influence the noise, delay, power dissipation and speed of circuits. Earlier, the choice of material suitable for interconnects was based on its conductivity and adherence on SiO_2. However, at high frequencies, certain problems like skin effect, signal integrity and crosstalk-induced propagation delay become prominent and difficult to handle by traditional interconnects. Hence, the interconnect problem is turning acute, and the limits of conventional interconnects are approaching fast. These tribulations led the researchers to contemplate upon novel options for interconnect, which can be on par with the technology advancement. The quest for innovation led researchers comes up with ambitious concepts involving ***graphene-based interconnects*** (Murali et al., 2009), ***optical interconnects*** (O'Connor and Gaffiot, 2004), ***organic interconnects*** (Varnava, 2020), ***spintronic switches*** (Naeemi et al., 2016) and alternative metals as well as alloys and composites.

Prior to excavating into the detailed discussion on state-of-the-art interconnects, a brief overview of practical constraint and imminent challenges encountered by existing interconnecting wire technology is presented in the subsequent sections.

2.4.1 Limitations of Conventional (Al/Cu) Interconnect Technology

To get high-speed integrated circuits, it is a must to bring out interconnections that would allow a rapid transmission of information (Meindl, 2003). Earlier, aluminum was preferred to make metallic on-chip wires due to its lower resistivity, cost effectiveness and silicon compatibility. As device dimensions are reduced, the current density increases. This results in reduced reliability of VLSI circuits due to the electro migration and hillock formation, causing electrical shorts between successive levels of Al. To reduce the electro migration problem, two or more metal layers are used in the same level of interconnections. Some of the multilayer interconnect materials are Al/Ti/Cu, Al/Ta/Al, Al/Ni, Al/Cr, Al/Mg and Al/Ti/Si (Goel, 2007). One of the useful methods of reducing the hillocks on silicon-based circuits is to deposit a film of WSi or MoSi between Al and Si substrate. Complete elimination of hillocks was reported when the VLSI interconnections were fabricated by alternatively layering the Al and a refractory metal (Ti or W) layers. Other potential options for metallic conductors with electrical resistivity lower than aluminum were silver, copper and gold having bulk resistivity values as 1.6, 1.7 and 2.4 $\mu\Omega$ cm, respectively. Gold (Au) exhibits greater resistance towards electro migration but in terms of resistivity it shows meek improvement. Besides this, one of the major drawbacks of using gold is that Au due to diffusion with Si, creates deep levels in the bandgap and henceforth affects the electronic properties of a device. On the same track, yet another possible metal is silver (Ag), which offers low resistance to electro migration because of its low melting point. Although Ag is low resistivity but it also continuously diffuses in SiO_2 and creates deep levels in the bandgap. Moreover, building of a diffusion barrier for silver is very trivial.

Thus, because of aforementioned reasons, Al which was once the cost effective and preferred interconnect material for several technology generations was eventually replaced by Cu. In 1997, IBM announced plans to replace Al with Cu, a metal with lower resistivity and better reliability than pure Al (Koester et al., 2008). This material paradigm shifting transition in late 1990s was a groundbreaking achievement in VLSI industry as it opens new potential pathways for on-chip interconnects wire structures. In terms of electrical resistivity, Cu has proven itself to be a superior contender than Al. copper provides high current density (10^6 A/cm^2) leading to the electro migration effect being less significant. Until early 2000, copper is unanimously preferred interconnect material in industries. But in mid-2000, with the advent of nanometer era, Cu gradually started facing the problem of an increase in resistivity due to the phenomenon of surface roughness and grain boundary scattering. Unfortunately, at ultra-deep submicron range, the current density (J) of interconnect wires exceed the maximum limit of current that can be carried by copper conductors (6×10^6 A/cm^2) due to electro migration (Li et al., 2004). Moreover, Cu is hazardous for Si circuits as it quickly diffuses into the gate, source and drain regions of transistors that are fabricated on silicon. This diffusion harnesses the functionality of the transistor. To overcome this limitation, a low-k dielectric material had been introduced as a new fabrication technique, but is an expensive one. Soon it was also observed that Cu/low-k interconnect technologies for the nanometer range is impacting system performance through increased crosstalk, signal transmission delay and power dissipation (Wong, 2012). Cu/low-k interconnects confronted several materials as well as electrical issues on the nanometre scale. The material issues include increments in resistivity due to the presence of a highly resistive barrier layer and increased scattering of electrons with scaling at the surface and grain boundaries. The second issue is the lower conductivity of low-k dielectric materials than SiO_2. The higher current density demand from interconnects increases with the technology scaling. All these three reasons are resulting in a rise of interconnect temperatures at global wiring. Moreover, thermal conductivity slumps while the backend metal temperature (T_m) increases above the junction temperature. A hike in temperature along with scaling in Cu interconnecting lines with low-k dielectric introduces reliability issues commonly termed time dependent dielectric breakdown (TDDB). Electro migration (EM) lifetime of Cu wires, which is a quadratic function of current density (J) and exponential function of T_m, is negatively affected by aforementioned factors. Global interconnect latency degrades with technology diminution. At higher frequencies, the effective resistance value of wires also rises due to the skin effect in copper conductors. These electrical issues limit the accuracy of parasitic extraction in Cu/Low-k interconnects. Repeater insertion can be a solution for minimizing the delay but the repeaters further increase the overall power dissipation of the chip (Khursheed and Khare, 2020). However, all of these short-term goals to resolve the Cu-foreseen issues might seem to be impractical or just economically unfeasible.

The looming crisis demands the future interconnects with lower parasitic, lower electro migration, less process variability and stress effects at smaller interconnect dimensions. To clinch these demands and to bridge the gap, designers came up with novel solutions. A detailed review on such emerging state-of-the-art interconnects is carried out in the following section.

2.4.2 State-of-the-Art Emerging Interconnect Technology

a Optical Interconnect

Optics technology offers prospective advantages because of its extremely high carrier frequency, under the influence of a low-loss waveguide or free-space medium; it demonstrates a velocity near the speed of light with negligible energy dissipation within the waveguide. Moreover, optical interconnects can attain a greater bandwidth by adopting the WDM technique (Pepeljugoski et al., 2010). Optical interconnects based on optic technology are at variance from the electrical-based interconnect technology (Cu, CNT and GNR) mainly on the basis of facts; firstly, it offers very high latency and on top of it the whole power consumption is observed in the end devices rather than in the transmission media. Secondly, the nature of energy dissipation is predominantly static rather than dynamic. Thus, it can be concluded that if these mentioned facts are coupled with favorable interconnect architectures, it showcases novel opportunities for optical wires as interconnects or vias.

Even though optical interconnects seem to be a viable contender for state of art interconnects, yet the drawbacks of optics technology, like comparatively greater transmission medium pitch, overshadow its chances of being the superior interconnect technology. Researchers try to summon this issue by adopting the technique of wavelength division multiplexing (WDM) available for optical wires. Furthermore, the scaling of wire dimensions increases the complexity of WDM, which limits its use only for modeling global and intermediate/semi-global interconnects. In addition to this, it is necessary to employ extremely efficient and easily manufacturable end devices. End devices like optical modulators and detectors with superior performance are necessary for super fast speed, small loss and ultra-low-power optical to electrical conversion and vice a versa. In order to achieve perfect electrical-to-optical conversion, the quantum confinement stark effect (QCSE) of III–V quantum well modulator is needed for providing fast and efficient modulation (Dokania, 2009). Unfortunately, it suffers from a lack of silicon compatibility. Recently, a Si/SiGe quantum well modulator has been demonstrated, providing a high potential for silicon-compatible manufacturability. In addition to the quantum well-structured modulator, a Si ring resonant-type modulator also offers promising performance; however, its vulnerability of high Q (quality factor) to temperature and process variations needs to be resolved. For optical-to-electrical conversion operation, a photo-detector with Ge on Si substrate can provide high speed with low parasitic capacitance and coupling loss.

b Spintronic Switches-Based Interconnect

To resolve the previous limitation, many researchers suggested certain interconnect design frameworks by using nano-scale spin torque switches designed for ultra-low-power sub-threshold wire modeling. In these techniques, the data signal is transmitted in the form of current pulses, with amplitudes in the order of a few micro-amperes that flow across a small terminal voltage of less than 50 mV. In order to receive and convert a high-speed current mode signal into binary voltage levels, sub-nanosecond spin-torque switching of scaled nano-magnets can be used. Magnetic tunnel junction (MTJ) converts a current mode signal into binary voltage levels using a CMOS inverter. Literature review explains that the spin polarity of nano-scale magnets can be flipped at a sub-nanosecond speed using small charge currents (Lim, 2004; Ngo et al., 2011). Due to low voltage and low-current signaling with minimal signal-conversion overhead, the spintronics-based interconnect offers extremely compact and simplified designs for variable high frequency on chip interconnects. Magneto-metallic spin-torque devices such as a domain wall magnet and spin valves acts as ideal receivers for current-mode signals, this is primarily due to their small resistance and probability of low-current, high-speed switching. These types of low-resistance receiver ports allow ultra-low-voltage biasing of interconnect networks. As a result of this, a notable reduction in static power consumption is observed due to current-mode signaling.

c Graphene-Based (CNT and GNR) Interconnect

To foreshadow the discussion to come, it is worthwhile stating in advance that among the various physically available allotropes of carbon, graphene is proven to be the best for on-chip nano-electronic interconnect applications (Van Noorden, 2006), as it is the strongest and thinnest material manifesting unique thermal and electronic traits. Graphene has a 2D hexagonal framework and is a basic building block for CNTs and GNRs. Both CNTs and GNRs, because of their large mean free path, commendable thermal conductivity and current-carrying capabilities can serve as backbones for next-generation on-chip interconnects. These unique electrical, thermal and mechanical properties of graphene are because of the atomic arrangements of carbon atoms, the band structure and crystal lattice. As literature review suggests, graphene is an ideal entrant for engineering novel materials. The all surface nature of graphene opens the door of opportunities for tailoring its properties by techniques of surface treatments like chemical functionalizing. For instance, graphene can be transformed into a bandgap semiconductor like hydrogenated graphene or into an insulator like fluorographene. Moreover, graphene flakes can be kept into dispersions, thereby many of its exceptional properties are retained and are utilized in order to realize composite materials with enhanced performance.

It can be stated that graphene is not only unique due to its outstanding properties but also due to its paradigm for a new category of materials, which will eventually evolve depending on the rise of technology based on graphene.

The purpose of the following section is to make the reader cognizant of the properties of electrons in CNTs and GNR. To effectuate the purpose, it is essential to first familiarize oneself with the mathematical techniques and physical ideas behind the theory of electrons, particularly in solids. The brief discussion on the behavior of electrons in solids will help in developing intuition by considering an introductory quantum mechanical description of electrons, and subsequently exploring the crystal structure.

The inkling developed and the understanding gained will be put to work in the development of the band structure of CNTs and graphene nanoribbons. References are scattered throughout the chapter for readers who want a detailed coverage of the basic concepts.

The assurance given by graphene to complement or replace semiconductors in the case of nanoelectonics is governed by various factors like its 2D nature, ease of processing and direct control of charge carriers and quasi-relativistic i.e. quick-moving electronic excitations resulting in a high μ (equal between electrons and holes). The electronic band structure of graphene is of primary importance because (i) it is the starting point for the understanding of graphene's solid-state properties and analysis of graphene devices (CNTFET and GNRFET) and (ii) it is also the starting point for the understanding and derivation of the band structure of CNTs. We begin by broadly discussing the general quantum mechanical wave nature of electrons and also the crystal structure or lattice (i.e. periodic arrangement of atoms in crystalline solid matter).

Behavior of Electrons in Empty Space

The rudimentary model of an electron is the model of a free electron in empty space. *Free electron* is the term coined for electrons that have no forces or potential acting on them. This didactic model of a free electron is used to determine a energy-wave vector relationship. In a 1D space measured by variable $\boldsymbol{x}$, the Schrödinger equation is

$$H\,\psi(k, x) = E(k)\psi(k, x) \tag{2.1}$$

$$\left(-\frac{h^2}{2m}\frac{\partial^2}{\partial x^2} + U(x)\right)\psi(k, x) = E(k)\psi(k, x) \tag{2.2}$$

where $\boldsymbol{H}$ is the Hamiltonian operator, $\boldsymbol{E}$ is the electron's total energy, $\boldsymbol{\psi}$ is the electron wavefunction, $\boldsymbol{k}$ is the wave vector, $\boldsymbol{h}$ is Planck's constant, $\boldsymbol{m}$ is the mass of electrons and $\boldsymbol{U}(\boldsymbol{x}) = 0$ is for all $\boldsymbol{x}$ in a free- electron model.

This is a differential equation with a general solution in the form of exponentials:

$$\psi(k, x) = Ae^{ikx} \tag{2.3}$$

Two relations are used in order to deduce the energy of electrons: the first one is from classical mechanics, which treats the electron as a particle with energy expressed in terms of its momentum $\boldsymbol{p}$:

$$\boldsymbol{E} = \frac{\boldsymbol{p}^2}{2\boldsymbol{m}}, \tag{2.4}$$

The second relation is the de Broglie wave-duality postulate, which assigns wave-like behavior to a particle with momentum $\boldsymbol{p}$. Thus, substituting $\boldsymbol{p} = \hbar \boldsymbol{k}$ into Eq. 2.4, the wave vector relationship for a free electron in free space leads to the parabolic dispersion:

$$\boldsymbol{E} = \frac{\boldsymbol{h}^2\boldsymbol{k}^2}{2\boldsymbol{m}} \tag{2.5}$$

For a 3D space, the wave vector $\boldsymbol{k}$ is replaced by the 3D wave vector $\boldsymbol{K}$, such that $\boldsymbol{K}^2 = \boldsymbol{k}_x^2 + \boldsymbol{k}_y^2 + \boldsymbol{k}_z^2$.

Thus, we may conclude that free electrons in space can oscillate as plane waves with a continuous set of wave vectors and, correspondingly, a continuous energy

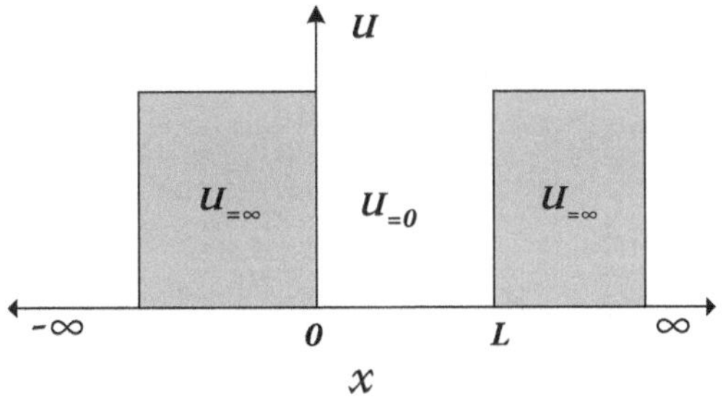

FIGURE 2.5 Potential profile illustrating a 1D finite empty solid of length L.

spectrum. Based on this conclusion, we will study the behavior of electrons in a finite empty solid.

Behavior of Electrons in a Finite Empty Solid

Let us assume the empty infinite solid of the previous section has an empty finite solid of length L with a potential shown in Figure 2.5 and can be mathematically expressed in 1D as

$$U(x)\begin{cases} 0 \ for \ \ 0 < x < L \\ \ \infty \ elsewhere. \end{cases} \tag{2.6}$$

For analysis, substitute the potential given in Eq. 2.6 in the Schrödinger equation. The general solution is given by complex exponential functions indicative of an oscillating wave:

$$\psi(k, x) = Ae^{ikx} + Be^{-ikx} \tag{2.7}$$

The finite size of a solid imposes a boundary condition (i.e. the electron cannot escape from the solid since the infinite potential exerts an infinite force that binds the electron) that places a restriction on the waveform.

$$\psi(x = 0) = \psi(x = L) = 0. \tag{2.8}$$

On applying a boundary condition at $\boldsymbol{x} = 0$, this requires B = –A and simplifies the wave function, $\boldsymbol{\psi(k, x)} = \boldsymbol{C}$ sin $\boldsymbol{kx}$, where $\boldsymbol{C}$ is a complex constant (with a magnitude $|C| = \sqrt{2/L}$). The boundary condition at $\boldsymbol{x} = \boldsymbol{L}$ requires

$$k = \frac{n\pi}{L}, \quad n = 1, 2, 3 \ldots \tag{2.9}$$

n is a quantum number that indexes the allowed wave vectors. On inputting the values for the wave vector into Eq. 2.5, the energies obtained are

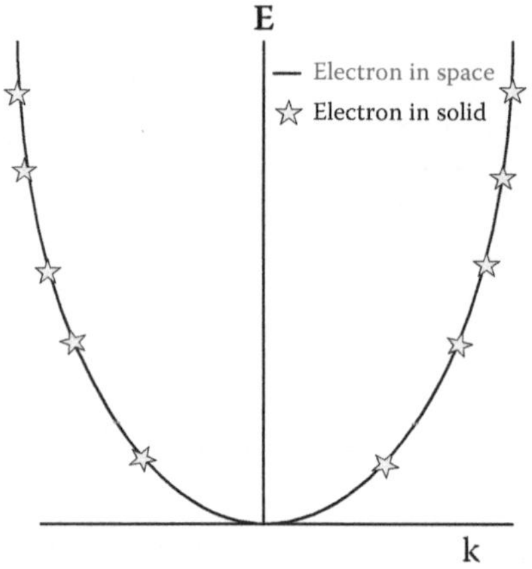

FIGURE 2.6 Energy-wave vector dispersions of a free electron in space and in a finite solid.

$$E = \frac{h^2k^2}{2m} = \frac{h^2\pi^2}{2mL^2}n^2. \tag{2.10}$$

Thus, from the above analysis, we may conclude that

1. *Dispersion quantization:* Due to finite size of a solid, the electron wave vector and energy can only take discrete values.
2. *Zero-point energy:* The energy of the electron cannot vanish. The lowest possible energy, called the zero-point energy, occurs when $n = 1$. The dispersion is illustrated in Figure 2.6.

The dispersion can be extended to a 3D empty finite solid by replacing the wave vector $\boldsymbol{k}$ with the 3D wave vector $\boldsymbol{K}$, which leads to three independent quantum numbers, n_x, n_y, n_z, representing the boundary conditions from each spatial dimension. The resulting 3D dispersion is

$$\boldsymbol{E} = \frac{\boldsymbol{h^2k^2}}{\boldsymbol{2m}} = \frac{\boldsymbol{h^2\pi^2}}{\boldsymbol{2mL^2}}\boldsymbol{n_x^2} + \boldsymbol{n_y^2} + \boldsymbol{n_z^2} \tag{2.11}$$

Similar to 1D dispersion, the 3D dispersion has the same physics but with the addition to the concept of *degeneracy* (which means a different set of quantum numbers results in the same energy). The parabolic dispersion obtained is used to predict and explain many of the properties of electrons from 3D solids, such as bulk semiconductors to 1D molecular solids such as CNTs.

Behavior of Electrons in a Periodic Solid: Kronig–Penney Model

Aforementioned models discussed were based on the behavior of an electron in an infinite and finite empty solid. However, real crystalline solids have a periodic arrangement of atoms that exerts a force on electrons. Therefore, in reality, any electron in a solid cannot be completely free. In the pursuit of spatial freedom, an electron can, at best, be nearly free.

Thus, now we study the behavior of electrons by modeling the periodic potential field in the solid or crystal lattice, based on the Kronig–Penney model (Kronig and Penney, 1931).

By Bloch's theorem, the solution of differential equation Eq. 2.12 is given by Eq. 2.13:

$$\frac{d^2\psi}{dx^2} + \frac{2m}{h^2} + (E - U(x))\psi = 0 \tag{2.12}$$

$$\psi(x) = u(x)e^{ikx} \tag{2.13}$$

where $u(x)$ is an amplitude modulating function with the periodicity of the lattice potential, $u(x) = u(x + a)$. Bloch's theorem states that the wave function is periodic with the length of the lattice $\psi(0) = \psi(L)$, resulting in quantization of the wave vector:

$$e^{ikL} = 1 \rightarrow k = \frac{2\pi}{L}n, \tag{2.14}$$

To find out $u(x)$, take a unit cell and examine the space between the potentials when $U(x) = 0$; Schrödinger's equation reduces to

$$\frac{d^2\psi}{dx^2} + \gamma^2\psi = 0, \quad \gamma^2 = \frac{2mE}{h^2} \tag{2.15}$$

Substituting the values of Eq. 2.13 into Eq. 2.15 yields

$$\frac{d^2u}{dx^2} + 2ik\frac{du}{dx} + (\gamma^2 - k^2)u = 0, \tag{2.16}$$

The solution of this standard second-order linear differential equation with constant coefficients is given by Eq. 2.17:

$$u(x) = (A \cos\gamma x + B \sin\gamma x)e^{-ikx} \tag{2.17}$$

Applying periodic boundary conditions for continuity of $u(x)$ and its slope

$$u(x) = u(a). \tag{2.18}$$

On integrating the Schrödinger equation over a very small region 2ε about $x = 0$, we derive the appropriate boundary condition for the slope

$$-\frac{h^2}{2m}\int_{-\varepsilon}^{\varepsilon}\frac{d^2\psi}{dx^2}dx+\int_{-\varepsilon}^{\varepsilon}C\delta(x)\psi dx=E\int_{-\varepsilon}^{\varepsilon}\psi dx \tag{2.19}$$

On evaluating the integral assuming ψ on RHS to be constant for an infinitesimal interval, we get:

$$\psi'(\varepsilon^+)-\psi'(\varepsilon^-)=\frac{2mC}{h^2}\psi(0) \tag{2.20}$$

Now from Eq. 2.13, on substituting the value for ψ and simplifying we get

$$u'(\varepsilon^+)-u'(\varepsilon^-)+ik(u(\varepsilon^+)-u(\varepsilon^-))=\frac{2mC}{h^2}u(0). \tag{2.21}$$

Since $u(x)$ is a periodic with length a, hence we can replace $x=\varepsilon^+$ with $x=0$; and $x=\varepsilon^-$ with $x=a$.

$$u'(0)=u'(a)+\frac{2mC}{h^2}u(0) \tag{2.22}$$

On simplifying Eqs. 2.17, 2.18, and 2.22, we get two simultaneous equations:

$$A(1-e^{-ika}cos\gamma a)=Be^{-ika}sin\gamma a \tag{2.23a}$$

$$A\left(ike^{-ika}cos\gamma a+\gamma e^{-ika}sin\gamma a-ik-\frac{2m}{h^2}C\right)=B(\gamma e^{-ika}cos\gamma a-ike^{-ika}sin\gamma a-\gamma). \tag{2.23b}$$

$$\cos ka=cos\gamma a+P\frac{sin\gamma a}{\gamma a}. \tag{2.24}$$

where $P=maC/h^2$.

With the help of Eq. 2.24 obtained from the Kronig–Penney model, we can plot the dispersion curve shown in Figure 2.7.

Some of the key features of the dispersion plot in Figure 2.7 are briefly discussed below:

- ***Brillouin Zone***: Eq. 2.24 does not have a unique solution. For every allowed energy that yields an LHS value within ±1, there will be a corresponding infinite set of wave vectors *k*. The space or zone restricted to these *k*-values is formally called the first ***Brillouin zone*** and is illustrated as the shaded region in Figure 2.7.
- ***Nearly Free Electron***: In the limit that P → 0, then k → λ, we get the dispersion for an electron in an empty solid of size *L*:

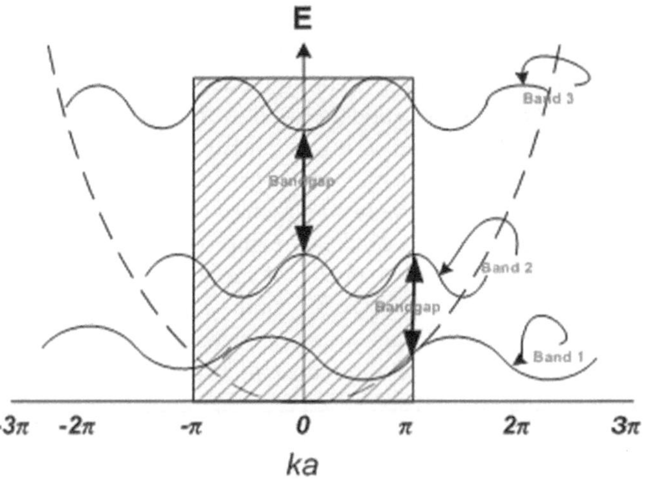

FIGURE 2.7 Kronig–Penney model based on electron dispersion curve showing allowed bands separated by bandgaps. The shaded rectangular portion of the curve shows dispersion of a free electron.

$$E = \frac{h^2k^2}{2m} = \frac{h^2\pi^2}{2mL^2}n^2 \tag{2.25}$$

Thus, in the limit of a weak periodic potential, the electron behaves as if it were nearly free; it can move around in the solid with negligible or no impediment from the periodic potential. The nearly free electron model is particularly useful in predicting fairly accurate band structures of bulk metals and the existence of bandgaps in crystalline solids.

- ***Tightly Bound Electron:*** In the opposite limit, when P → ∞, then the solutions to Eq. 2.24 exist only when $\gamma a = n\pi$, and the dispersion becomes

$$E = \frac{h^2k^2}{2m} = \frac{h^2\pi^2}{2ma^2}n^2 \tag{2.26}$$

This informs us that, in the limit of an increasingly strong periodic potential, the electron behaves as if it were tightly bound by the strong potential, which in reality represents the attraction from the atomic nucleus. The case of a strong periodic potential is of significant practical interest and is formally treated by methods such as the *tight-binding model*. In fact, this is the most widely used model for accurately describing the band structure of graphene and carbon nanotubes. This model will be discussed in more detail in the succeeding sections on derivation of the band structure of graphene.

Before moving ahead, an important thing to be noted is that as we view the electrons strictly as waves, then wave principles, such as reflections, are applicable. It is the reflections of electron waves from the potential wall that lead to *bandgaps*. And, analogous to electromagnetic waves, these reflections (commonly known as

Bragg reflections in solid-state physics) are an integral multiple of wavelength, and that is why we observe the bandgaps occurring at $ka = \pm n\pi$.

Although the pedagogical Kronig–Penney model gives much insight into how electrons behave in a crystalline solid (a solid with a periodic potential), it has certain limitations. First and foremost, it is a general model that does not take into account the specific lattice arrangement of a particular solid of interest. For instance, if we desire to determine the band structure of nanotubes and nanoribbons, the Kronig–Penney model is utterly inadequate in providing basic information, such as the bandgap dependency on the diameter of these nanostructures.

Thus, in order to determine the specific band structure of a particular solid, a thorough understanding of the crystal structure of that particular solid is must. The study of a crystal structure can be bifurcated in two domains. One domain is in position (sometimes called direct) space and the other domain is in reciprocal space. Although a brief discussion on both of the domains is given in the manuscript, the prime focus will be on the discussion of the crystal structure that is used for the development of two-dimensional (2D) or reduced-dimensional solids (graphene and CNTs).

For the said purpose, a systematic mathematical formalism is needed to describe the geometry of the lattice. From this mathematical formalism, we can then develop a technique to compute the band structure of particular solids. In general, the logical relation that is the foundation of crystal structure is the combination of Bravais lattice (a technical name for the fundamental irreducible microscopic arrangements of points) and the Basis (an integer number that represents how many atoms are attached to each Bravais lattice point).

Crystal Lattice = Bravais Lattice + Basis

Bravais Lattice

Bravais lattice, which is named after the French physicist Auguste Bravais, who in the mid-1800s pointed out that there are 14 unique fundamental lattices in 3D space, is the most fundamental way to describe a crystal lattice. Bravais lattice is also known as direct lattice.

In simple words, Bravais lattice can be defined as an array of discrete points with an arrangement and orientation that look exactly the same from any of the discrete points; that is, the lattice points are indistinguishable from one another. To make this statement more clear, let us look at a pictorial example, shown in Figure 2.8.

From Figure 2.8(a), we infer that since every discrete point in the 2D rectangular lattice sees exactly the same environment, thus the rectangular lattice is a Bravais lattice and has a basis of one, unlike the case of honeycomb lattice shown in Figure 2.8(b). For one, point A is directly looking at a single lattice point, while point B faces two lattice points at angles of ±60°. Consequently, the honeycomb, as it is, is not a Bravais or fundamental lattice. However, the honeycomb crystal is still a lattice because it consists of repeating unit cells and, as a result, it can be converted or mapped to a Bravais lattice. For the particular example of a honeycomb,

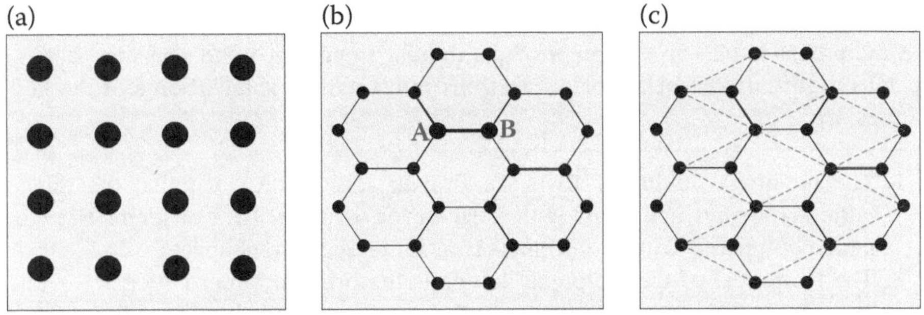

FIGURE 2.8 (a) Rectangular lattice is a type of Bravais lattice; (b) honeycomb lattice is a crystal lattice but not a Bravais lattice; (c) honeycomb lattice that can be converted to a Bravais lattice (with a basis of two) represented by dashed lines.

consider regrouping points A and B together (and all repeating instances of A and B); the resulting lattice is shown in Figure 2.8(c), which is now a Bravais lattice, and has a basis of two. That is, every Bravais lattice point represents two crystal lattice points. In crystalline solids, the real points are locations of atoms and the basis is conveniently interpreted as the number of atoms grouped together at each Bravais lattice point. It is important to keep in mind that the Bravais lattice is not always the same as the crystal lattice.

An outstanding property of crystal lattices that we will later employ in the derivation of electronic band structures is the nearest neighbors. The nearest neighbors are the number of points that are equally closest to a given point in the Bravais lattice.

Primitive Vectors

To describe the position of Bravais lattice points, we need to define a coordinate origin and vectors ($\mathbf{a_1}$ and $\mathbf{a_2}$ in 2D space). These coordinate vectors associated with the Bravais lattice are called the *primitive vectors*. Any integral multiple of these vectors must arrive at a Bravais lattice point, i.e. the position vectors of any lattice point are $R = n_1 a_1 + n_2 a_2$, where n_1 and n_2 are positive or negative integers. To state it in a different manner, the entire Bravais lattice can be constructed by translating the coordinate vectors in integral steps throughout space.

Primitive Unit Cell

Primitive unit cell is the most basic unit cell of the Bravais lattice. The primitive unit cell has two distinguishing properties: (1) it is a region of space that contains exactly one Bravais lattice point; (2) it re-creates the lattice when translated through all the Bravais lattice vectors without leaving gaps or generating overlaps. All primitive unit cells occupy exactly the same area or volume.

Reciprocal Lattice

The reciprocal lattice is a concept of paramount importance and provides fundamental insight into the behavior of electrons in crystalline solids. Some of the key features are:

1. The reciprocal lattice is always a Bravais lattice, provided that the direct lattice is always a Bravais lattice. However, it might be a different Bravais lattice compared with the original Bravais lattice in real space.
2. The reciprocal of the reciprocal lattice is the original direct lattice.
3. The direct lattice is measured in terms of a position vector R with units of length; the reciprocal lattice is measured in terms of the wave vector K with units of 1/length.
4. The reciprocal lattice satisfies the basic relation $e^{-iKR} = 1$ where K is the set of wave vectors that determine the sites of the reciprocal lattice points and R is the Bravais lattice position vector as usual. The basic relation originates from the Fourier analysis of the direct lattice.
5. If s is the area of the direct lattice primitive cell, then $(2\pi)^2/s$ is the area of the reciprocal lattice primitive cell.
6. The direct lattice exists in real space or position space; the reciprocal lattice exists in reciprocal space, which is sometimes called the Fourier space, momentum space or simply k-space. This is because the crystal momentum is directly proportional to the wave vector.

Band Structure of Graphene The motive of this section is to describe the physical and electronic structure of graphene. We begin by briefly discussing carbon and then swiftly focus on graphene.

Carbon is a Group IV element, active in producing many molecular compounds and crystalline solids. Carbon has four valence electrons, which tend to interact with each other to produce the various types of carbon allotropes described in Table 2.1.

In its elemental form, the four valence electrons occupy the $2s$ and $2p$ orbitals. When carbon atoms come together to form a crystal, one of the $2s$ electrons is excited to the $2p_z$ orbital, resulting in the formation of hybridized orbits. Graphene is a planar allotrope of carbon where all the carbon atoms form covalent bonds in a single plane. While forming graphene, three atomic orbitals of the carbon atom, $2s$,

TABLE 2.1
Allotropes of carbon

Dimension	0D	1D	2D	3D
Allotrope	C60 buckyball	Carbon nanotubes	Graphene	Graphite
Structure	Spherical	Cylindrical	Planar	Stacked planar
Hybridization	sp^2	sp^2	sp^2	sp^2
Electronic properties	Semiconductor	Metal or semiconductor	Semi-metal	Metal

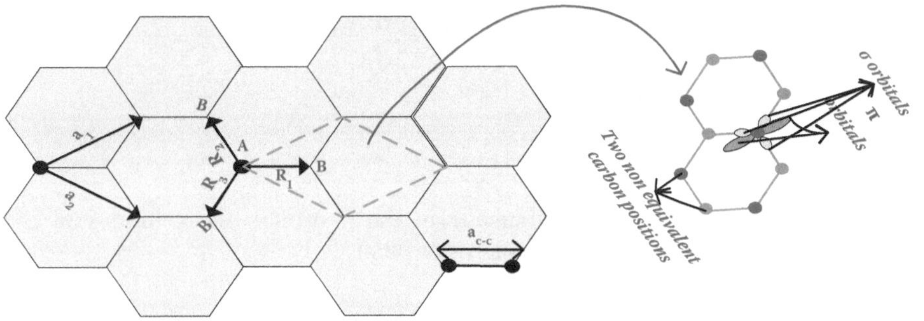

FIGURE 2.9 Graphene honeycomb lattice and illustration of orbital structure of graphene.

$2p_x$, and $2p_y$, are hybridized into three sp^2 orbitals. These sp^2 orbitals are in the same plane, while the remaining $2p_z$ is *perpendicular* to other orbitals as shown in Figure 2.9.

The sp^2 interactions result in three σ-bonds, which are the strongest type of covalent bond. The σ-bonds have the electrons localized along the plane, connecting carbon atoms and are responsible for the great strength and mechanical properties of graphene and CNTs. The $2p_z$ electrons form covalent bonds called π-bonds, where the electron cloud is distributed normal to the plane connecting carbon atoms. The $2p_z$ electrons are weakly bound to the nuclei and, hence, are relatively delocalized. These delocalized electrons are the ones responsible for the electronic properties of graphene and CNTs and as such will occupy much of our attention.

Graphene's honeycomb lattice shown in Figure 2.9 is typified as a Bravais lattice with a basis of two atoms. The carbon-carbon bond length, $a_{c-c} \approx 1.42\text{Å}$, is the primitive cell that can be described as an equilateral parallelogram with a side $a = \sqrt{3_{a_{c-c}}} = 2.46\text{Å}$. The unit vectors or lattice vectors of graphene are denoted by a_1 *and* a_2, whereas the reciprocal lattice vectors are denoted by b_1 *and* b_2 such that:

$$a_1 = \left(\frac{\sqrt{3}a}{2}, \frac{a}{2}\right), \quad a_2 = \left(\frac{\sqrt{3}a}{2}, \frac{a}{-2}\right) \text{ and } b_1 = \left(\frac{2\pi}{\sqrt{3}a}, \frac{2\pi}{a}\right),$$

$$b_2 = \left(\frac{2\pi}{\sqrt{3}a}, -\frac{2\pi}{a}\right)$$

where $|a_1| = |a_2| = a$ and $|b_1| = |b_2| = \frac{4\pi}{\sqrt{3}a}$. Each carbon atom is bonded to its three nearest neighboring atoms. The vector distance between carbon atom A and its neighboring carbon atom B is denoted by R_1, R_2, R_3, where $|R_1| = |R_2| = |R_3| = a_{c-c}$ and the value can be calculated as

$$R_1 = \left(\frac{a}{\sqrt{3}}, 0\right), \quad R_2 = -a_2 + R_1 = \left(-\frac{a}{2\sqrt{3}}, -\frac{a}{2}\right), \quad R_3 = -a_1 + R_1$$
$$= \left(-\frac{a}{2\sqrt{3}}, \frac{a}{2}\right)$$

To obtain the band structure of graphene in the π orbital, the solutions of the Schrödinger equation are required, which state that

$$H \ \psi(k, x) = E(k)\psi(k, x) \tag{2.27}$$

where H is the Hamiltonian operator, E is the electron total energy, ψ is the electron wave function and k is the wave vector. As discussed in preceding sections, because of the periodic nature of graphene, the total wave function can be obtained from a linear combination of the Bloch function u_i, using tight binding approximation. For each carbon atom, the u_i can be obtained from $2p_z$ orbitals of atoms A and B such that

$$\mathrm{u}_{\mathrm{A(B)}} = \frac{1}{\sqrt{\mathrm{N}}} \sum\nolimits_{\mathrm{A(B)}} e^{\mathrm{ik.r}_{\mathrm{A(B)}}} \ \mathrm{X}(\mathrm{r} - \mathrm{r}_{\mathrm{A(B)}}) \tag{2.28}$$

where $X(\mathrm{r})$ denotes the $2p_z$ orbital wave function for an isolated carbon atom. The electron wave function ψ is mathematically expressed as

$$\psi = C_A u_A + C_B u_B \tag{2.29}$$

The Schrödinger equation can be solved in matrix form after substituting a value of ψ from Eq. 2.29 into Eq. 2.27

$$\begin{pmatrix} H_{AA} & H_{AB} \\ H_{BA} & H_{BB} \end{pmatrix} \begin{pmatrix} C_A \\ C_B \end{pmatrix} = E \begin{pmatrix} S_{AA} & S_{AB} \\ S_{BA} & S_{BB} \end{pmatrix} \begin{pmatrix} C_A \\ C_B \end{pmatrix} \tag{2.30}$$

where $H_{ij} = u_i|H|u_j$ and $S_{ij} = u_i| \ u_j$; here, H_{ij} is the matrix elements of the Hamiliton or transfer integral and has units of energy, while S_{ij} is the overlap matrix element between Bloch functions and is unitless. Neglecting the overlap between $2p_z$ wave functions of different atoms can be considered as $S_{AB} = S_{BA} = 0$. Whereas for normalized case, the values taken are $S_{AA} = S_{BB} = 1$. Thus, the equation can be further simplified as

$$\begin{pmatrix} H_{AA} - E & H_{AB} \\ H_{BA} & H_{BB} - E \end{pmatrix} \begin{pmatrix} C_A \\ C_B \end{pmatrix} = \begin{pmatrix} 0 \\ 0 \end{pmatrix} \tag{2.31}$$

The matrix in Eq. 2.31 has a non-trivial solution if and only if

$$\begin{vmatrix} H_{AA} - E & H_{AB} \\ H_{BA} & H_{BB} - E \end{vmatrix} = 0$$

By virtue of symmetry of the graphene lattice, atoms A and B are not distinguishable; moreover, it is noted that $H_{AA} = H_{BB}$ and $H_{AB} = H_{BA}*$. Hence, the solution of the equation is given as

$$E = H_{AA} \mp |H_{AB}| \tag{2.32}$$

$$H_{AA} = \frac{1}{N} \sum \sum_{A+} e^{ik.(r_A - r_A^+)} \int X^+ (r - r_A) HX (r - r_A^*) d\tau \tag{2.33}$$

Taking into account the effects of nearest neighbors for each atom A(B) having three nearest neighbors B(A), the equation becomes (Javey and Kong, 2009)

$$H_{AA} = \int X^+ (r - r_A) HX (r - r_A^*) d\tau = E_0 \tag{2.34}$$

$$H_{AB} = \frac{1}{N} \sum_A \sum_B e^{ik.(r_A - r_B)} \int X^+ (r - r_A) HX (r - r_A^*) d\tau \tag{2.35}$$

$$H_{AB} = \frac{1}{N} \sum_i e^{ik.\rho_i}) \int X^* (r) HX (r - \rho_i) d\tau \tag{2.36}$$

with ρ_i as a vector connecting atom A to its three nearest neighbor B atoms. On further solving, keeping in mind the graphene coordinate system, we get the expression

$$H_{AB} \left(e^{ik.\rho_i} + e^{ik.\rho_2} + e^{ik.\rho_3}\right) \int X^* (r) HX (r - \rho_1) d\tau = \gamma_0 \left(e^{\frac{-ik_x a}{\sqrt{3}}} + 2 e^{\frac{-ik_x a}{2\sqrt{3}}} \cos\left(\frac{k_y a}{2}\right)\right) \tag{2.37}$$

From the tight-binding integral or transfer integral, γ_0 represents the strength of exchange interaction between nearest neighbor atoms. Thus, solving the previous equations, the expression for the energy dispersion can be expressed as

$$E = E_0 \mp \gamma_0 \left(1 + 4 \cos\left(\frac{\sqrt{3} k_x a}{2}\right) \cos\left(\frac{k_y a}{2}\right) + 4\cos^2\left(\frac{k_y a}{2}\right)\right)^{1/2} \tag{2.38}$$

Here in Eq. 2.38, the valence band of graphene formed by π orbitals is denoted by (−) sign, whereas (+) sign represents the conduction band produced by π^* orbitals. Figure 2.10 shows the plot of dispersion relation expressed by Eq. 2.38. The upper

(a)

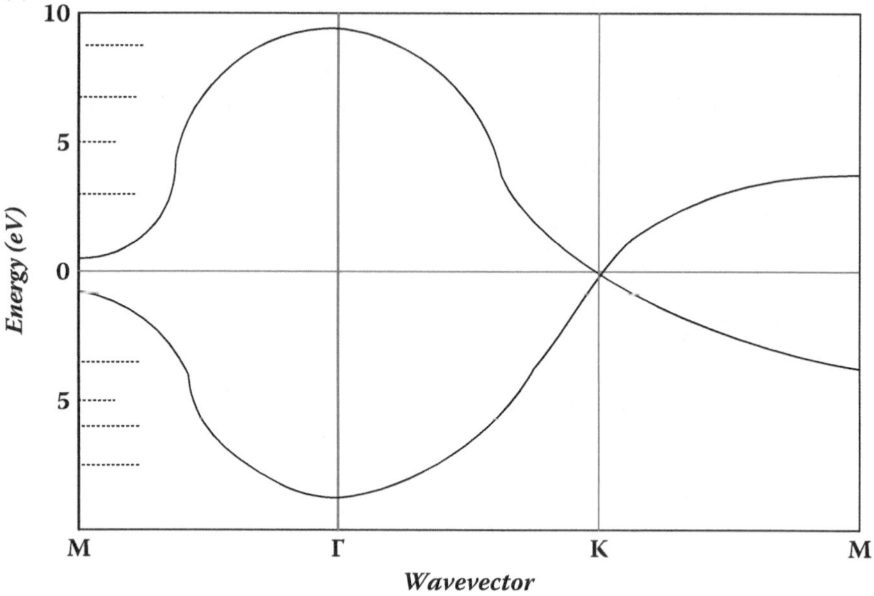

(b)

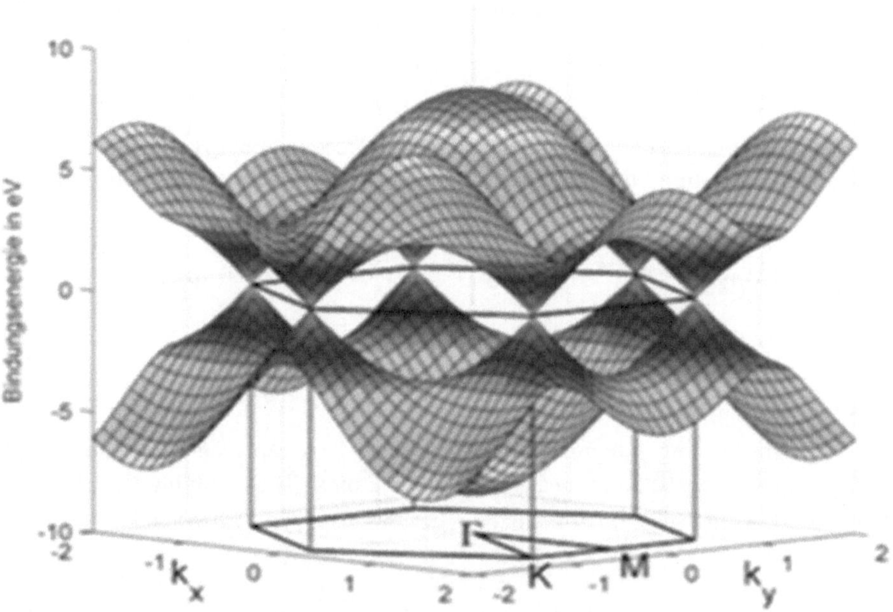

FIGURE 2.10 (a) Shows the plot of dispersion relation expressed by energy dispersion equation; (b) the nearest neighbor tight-binding band structure of graphene. The hexagonal Brillouin zone is superimposed and touches the energy bands at the K-points.

half of the dispersion is the conduction (π^*) band and the lower half is the valence (π) band. It is particularly of interest to identify the highest energy state within the valence band and the lowest energy state within the conduction band. These states occur at the K-point corresponding to $E = 0$ eV. Due to the absence of a bandgap at the Fermi energy, and the fact that the conduction and valence bands touch at E_F, graphene is considered a semi-metal or zero-bandgap semiconductor, in contrast to a regular metal, where E_F is typically in the conduction band, and a regular semiconductor, where E_F is located inside a finite bandgap.

Figure 2.10(b) represents the surface and contour plots of the energy dispersion, respectively. The six K-points at the corners of the Brillouin zone are the main feature of the energy dispersion of graphene. At these points, the conduction and valence bands meet, resulting in zero bandgap in graphene. These six K-points in graphene where the conduction and valence bands touch are frequently called the Dirac points. It can also be noted that the two K-points (K1 and K2) are non-equivalent due to symmetry. The circular contour around each K-point as illustrated in Figure 2.10(b) indicates the conic shape of dispersion near each K-point.

Band Structure of CNTs from Graphene A graphene sheet can be rolled up to form nanotubes in such a way that two atoms of carbon coincide. The manner in which the graphene sheets are wrapped into tubes uniquely defines the characteristics of obtained CNTs. In order to study the band structure of CNTs, an appropriate boundary condition is required. On assuming CNT to be a cylinder of infinite length with two associated vectors ($k_{\|}$ *and* $k_{\perp}$), it must satisfy a periodic boundary condition (i.e. the wave function repeats itself as it rotates 2π around a CNT). Here, the $k_{\|}$ the wave vector parallel to the axis of nanotubes is continuous in nature because of an infinitely long length of CNT and $k_{\perp}$ is the perpendicular wave vector along the circumference of nanotubes of diameter D.

$$k_{\perp}.\, C = \pi \;\; Dk_{\perp} = 2\pi m \tag{2.39}$$

The boundary condition leads to quantized values of allowed $k_{\perp}$ for nanotubes. The zone-folding scheme shown in Figure 2.11 is a powerful yet simple technique used for obtaining the band structure of CNTs. As illustrated in Figure 2.11, the 1D band structure of CNTs is acquired by cross-sectional cutting of energy dispersion of 2D graphene with the allowed $k_{\perp}$ states.

The 1D band structure of CNTs is ascertained by spacing in between the allowed $k_{\perp}$ states and their angles (obtained by wrapping indices) with respect to the surface Brillouin zone of graphene. In particular, the band structure near the Fermi level, most relevant for transport properties, is given by allowed $k_{\perp}$ states that are closest to the K-points. As depicted in Figure 2.11, when the allowed $k_{\perp}$ states passes directly through the K-points, the energy dispersion shows two linear bands crossing at the Fermi level without a bandgap. However, if the allowed $k_{\perp}$ states miss the K-points, then there are two parabolic 1D bands with an energy bandgap. Therefore, we can expect two different kinds of CNTs depending on the wrapping indices: metallic CNTs without a bandgap as shown in Figure 2.11(b), and semi-conducting CNTs with a bandgap as in Figure 2.11(c).

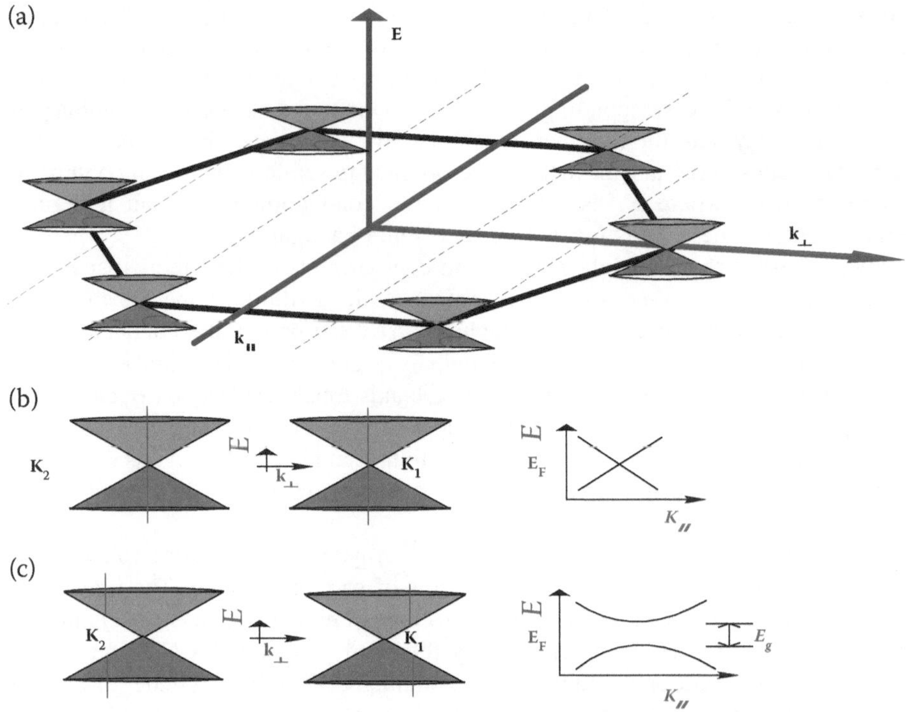

FIGURE 2.11 (a) A first Brillouin zone of graphene with conic energy dispersions at six K-points. The allowed k⊥ states in CNT are presented by dashed lines. The band structure of CNT is obtained by cross sections as indicated. Zoom-ups of the energy dispersion near one of the K-points are schematically shown along with the cross sections by allowed k⊥ states and resulting 1D energy dispersions for (b) a metallic CNT and (c) a semiconducting CNT.

Band Structure of Zigzag CNTs and the Derivation of the Bandgap Zigzag CNTs, due to their either metallic or semiconducting behavior, are the most attractive for researchers to explore. Moreover, they also possess high symmetry, leading to simple analytical expressions for many of the solid-state properties. The energy dispersion of zigzag CNTs can be obtained from the Brillouin zone wave vector by applying boundary conditions with a value of circumference $C = na$.

$$k_{\perp}.\ C = \pi\ Dk_{\perp} = 2\pi m = k_x n.\ a = 2\pi m \tag{2.40}$$

If integer n is a multiple of three (($n = 3q$; q *is an integer*), then there is an allowed k_x that coincides with a K-point, at (0,4π/3a). Further, solving it by substituting the value

$$k_x = \frac{2\pi m}{na} = \frac{3Km}{2n} = \frac{Km}{2q} \tag{2.41}$$

There is always an integer $m(=2q;\ q\ is\ an\ integer)$ that makes k_x pass through K-points so that these types of CNTs (with $n = 3q$; q *is an integer*) are always metallic without a bandgap as illustrated in Figure 2.11(b), whereas in two instances when n is not a multiple of 3. If $n = 3q + 1$, the k_x is closest to the K-point at $m = 2q + 1$.

$$k_x = \frac{2\pi m}{na} = \frac{3Km}{2n} = \frac{3K(2q-1)}{2(3q+1)} = k + \frac{K}{2}\frac{1}{3q+1} \tag{2.42}$$

In the same way, if $n = 3q - 1$, the k_x is closest to the K-point at $m = 2q - 1$. Thus,

$$k_x = \frac{2\pi m}{na} = \frac{3Km}{2n} = \frac{3K(2q-1)}{2(3q-1)} = k - \frac{K}{2}\frac{1}{3q-1} \tag{2.43}$$

In both the conditions, it is to be noted that the allowed k_x misses the K-point by

$$\Delta k_x = \frac{K}{2}\frac{1}{3q \pm 1} = \frac{2}{3}\frac{\pi}{na} = \frac{2}{3}\frac{\pi}{\pi D} = \frac{2}{3D} \tag{2.44}$$

Thus, from the Eq. 2.44, it can be concluded that the small misalignment between an allowed k_x and a K-point is inversely proportional to the diameter. Hence, the bandgap E_g is calculated from the slope of a cone near K-points is expressed as

$$E_g = 2 \times \frac{\partial E}{\partial k} \times \frac{2}{3D} = 2hv_F\left(\frac{2}{3D}\right) \approx 0.7eV/D(nm) \tag{2.45}$$

Hence, it can be stated from the inference drawn that the semiconducting CNTs with diameter $D = 0.8$–3 nm demonstrate a bandgap ranging from 0.2 eV to 0.9 eV. Based on the different values of p, two different cases occurs that describe the metallic and semiconducting properties of CNTs

i. p = 0; metallic with linear subbands crossing at the K-points.
ii. p = 1, 2; semiconducting with a bandgap $E_g \sim \frac{0.7eV}{D}(nm)$.

Applying same treatment to armchair CNTs leads to the conclusion that they are always metallic.

2.4.2.1 Dominion of CNT Interconnect

Carbon nanotubes have a long mean free path (several micrometers) and carry extremely high current densities than Cu wires (Raychowdhury and Roy, 2004). The CNTs with a diameter of a nanometer are grown in the form of seamless cylinders with the walls formed by one atomic layer of graphite (known as graphene).

Inevitably, a sound knowledge of the electronic and physical structure of graphene is a must to fully comprehend the behavior of electrons in CNTs. As discussed earlier, CNTs are obtained by technique of wrapping the graphene sheet governed by its chirality.

Before discussing how the notion of chirality is implemented in explaining the different configurations and band structure of CNTs, it is better to first give a brief introduction of the concept of chirality, as it is of fundamental importance and often unfamiliar to engineers.

Chirality is a property of asymmetry important in several branches of science. The word *chirality* is derived from the Greek χειρ, meaning "hand," a familiar chiral object. An object or a system is *chiral* if it is distinguishable from its mirror image; that is, it cannot be superimposed onto it, whereas achiral objects are types of objects that can be superimposable on its mirror image. For instance, a human hand is a chiral object and a circle is an achiral object.

Thus, comprehending the concept of chirality is a must as it helps in classifying the physical and electronic structure of CNTs. The CNTs that are not super imposable on their own mirror images are classified as chiral CNTs, whereas all other types of CNTs that cannot be superimposed are termed achiral CNTs (armchair CNTs or zigzag CNTs). Figure 2.12 shows sketches of different CNT structures (Scarselli et al., 2012).

Depending on axial helicity (chirality), the direction in which CNTs are rolled, they exhibit metallic or semi-conducting behavior. Semiconducting CNTs, with the help of a gate electrode, can be made to switch on/off and are thus preferred for

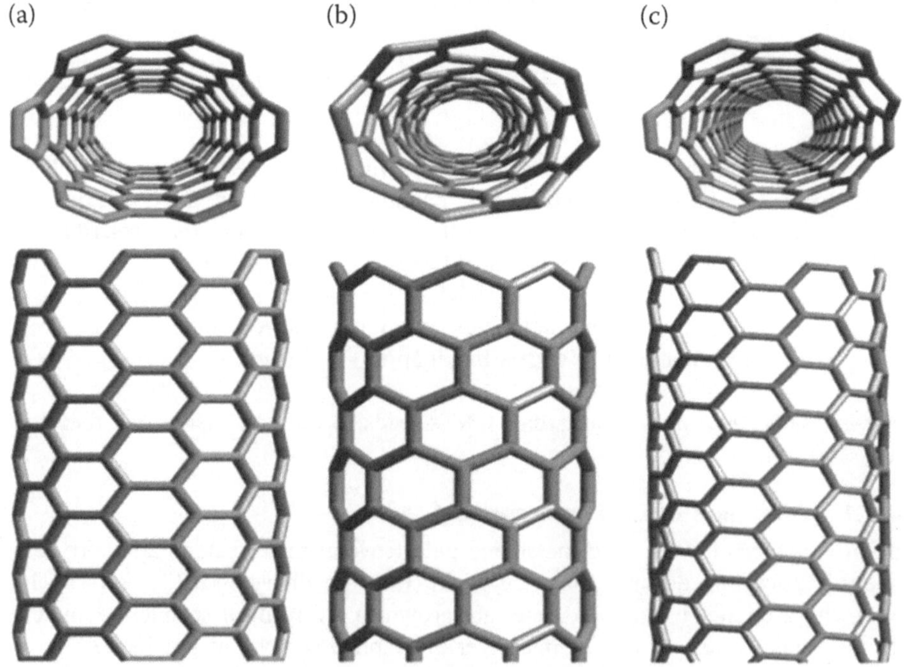

FIGURE 2.12 Sketches of three different SWCNT structures: (a) armchair nanotube; (b) zigzag nanotube and (c) chiral nanotube.

implementing field-effect transistors, whereas metallic carbon nanotubes, due to their extremely large current-carrying capability, are suitable for interconnect applications. Thus, current conduction in interconnect wires is due to metallic nanotubes only and not semiconducting ones (Naeemi and Meindl, 2005).

As aforementioned, compared to conventional materials, the carbon nanotubes evince remarkable mechanical and electrical properties, which are summarized in Tables 2.2 and 2.3 (Li et al., 2003).

The CNT Lattice

The astounding properties of CNTs emanate from the unusual electronic structure and symmetry of graphene. Furthermore, the sub-lattice structure of graphene engenders a unique property called pseudo-spin that is not seen in other materials

TABLE 2.2
Comparative chart of mechanical properties of CNT with other material

Material	Young's Modulus (T Pa)	Tensile Strength (G Pa)	Elongation at Break (%)	Thermal Conductivity (W/m k)
SWCNT	1-5	13-53	16	3,500-6,600
MWCNT	0.27-0.95	11-150	8.04-10.46	3000
Stainless steel	0.186-0.214	0.38-1.55	15-50	16
Kevlar	0.06-0.18	3.6-3.8	~2	~1
Copper	0.11-0.128	0.22		385
Silicon	0.185	7		149

TABLE 2.3
Comparative chart of electrical properties of CNT with other materials

Semiconductor					Metal		
Parameter	**Semiconducting SWCNT**	**Si**	**GaAs**	**Ge**	**Parameter**	**Metallic SWCNT**	**Copper**
Bandgap (eV)	0.9/diameter	1.12	1.424	0.66	Mean Free Path (nm)	1,000	40
Electron Mobility (cm^2/Vs)	20,000	1,500	8,500	3,900	Current density (A/cm^2)	10^{10}	10^6
Electron Phonon Mean Free Path (Å)	~700	76	58	105	Resistivity (Ω· m)	$\sim 10^{-5}$	1.68×10^{-8}

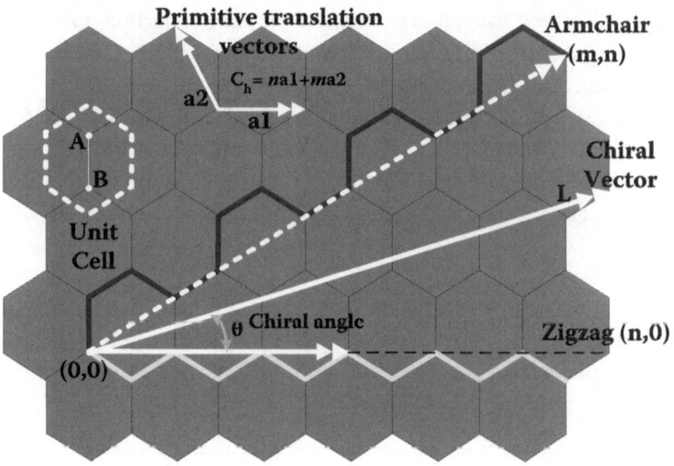

FIGURE 2.13 CNT geometry: illustrating the concept of constructing CNT from graphene.

(Aoki and Dresselhaus, 2014). The pseudo-spin flip reduces the charge carrier scattering. The graphene sheets are comprised of hybridized sp^2 carbon atoms patterned in a hexagonal arrangement. These hexagonal rings in CNT are joined coherently with neighboring carbon atoms and are equally spaced from one another (except at the edges). Despite this arrangement, CNT is more reactive than planar graphene because CNT is not purely sp^2 hybridized and some sp^3 hybridization is also found. The degree of sp^3 hybridization rises with a decrease in tube diameter, causing variable overlapping of energy bands. They have a bandgap in nearly all directions in k-space, but also have a vanishing bandgap in particular directions; thus, they are called zero-bandgap semiconductors. Wrapping the graphene sheet to obtain a nanotube, quantize the momentum of the electrons moving around the circumference of the tube (Satio et al., 1992). The graphene sheet can be wrapped in a variety of ways, as depicted in Figure 2.13.

As depicted in Figure 2.13, the honeycomb lattice of graphene and the primitive lattice vectors $\boldsymbol{a}_1$ and $\boldsymbol{a}_2$ are defined on a plane with unit vectors $\vec{x}$ and $\vec{y}$:

$$\boldsymbol{a}_1 = \left(\frac{\sqrt{3}}{2}; \frac{1}{2}\right)\boldsymbol{a}, \quad \boldsymbol{a}_2 = \left(\frac{\sqrt{3}}{2}; -\frac{1}{2}\right)\boldsymbol{a} \tag{2.46}$$

here, $\boldsymbol{a}$ is an underlying Bravais lattice constant $\boldsymbol{a} = \sqrt{3}\ \boldsymbol{a}_{c-c} \approx 2.46$ Å and $\boldsymbol{a}_{c-c} \approx 1.42$ Å. $\boldsymbol{a}_{c-c}$ is the carbon – carbon bond length.

In general, the CNT is characterized by three geometrical parameters: the chiral vector $\mathbf{C}_h$, the translational vector T and the chiral angle θ.

The chiral vector $\mathbf{C}_h$ is a geometrical parameter uniquely defined as the vector connecting any two primitive lattice points of graphene such that when folded into a nanotube these two points are coincidental or indistinguishable. Mathematically, it

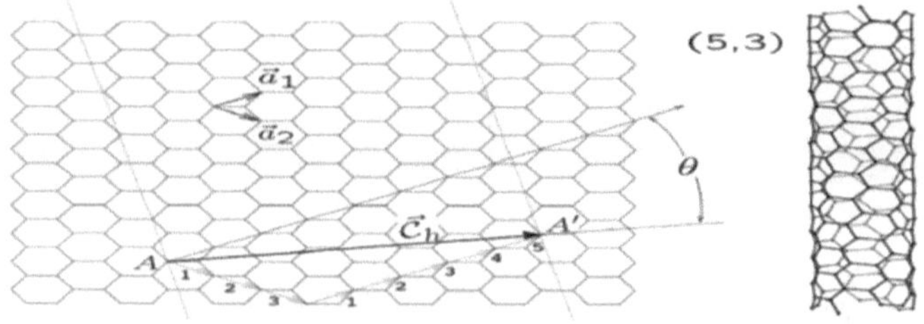

FIGURE 2.14 Graphene lattice. The lattice vectors a_1 and a_2. The chiral vector $C_h = 5a_1 + 3a_2 = (n, m)$ represents a possible wrapping of the two-dimensional graphene sheet into a tubular form.

can be expressed as $\mathbf{C}_h = na_1 + ma_2 = (n, m)$, ($n$, m are positive integers, $0 \le m \le n$) and the resulting nanotube is described as (n,m) CNT.

The m and n values for a particular CNT can be easily obtained by counting the number of hexagonal rings separating the margins of the $\mathbf{C}_h$ vector following a_1 first and then a_2, as demonstrated in Figure 2.14.

The direction perpendicular to $\mathbf{C}_h$ is the tube axis. The chiral angle θ is defined by the $\mathbf{C}_h$ vector and the $\mathbf{a}_1$ zigzag direction of the graphene lattice.

Based on (n, m) indices, CNTs can be grouped as armchair nanotubes ($n = m$), zigzag nanotubes ($n = 0$ *or* $m = 0$) and chiral nanotubes (any other combination) (Figure 2.15).

Furthermore, CNT is metallic when ($n - m = 3l$); l is an integer and is semiconducting when ($n - m \neq 3l$). The sheet of honeycomb lattice carbon molecules runs along the nanotube axis in armchair CNTs and around its circumference in zigzag CNTs. Zigzag CNTs have pseudo-energy gaps due to the molecular interactions among nanotubes that prevents free flow of current as these gaps destroy the rotational symmetry while armchair nanotubes conduct in bundled as well as in isolated variations. The researchers do not precisely know how much these pseudo-energy gaps slow current but theories suggest deterioration of conducting properties because of the emergence of these gaps.

The diameter d of a nanotube is procured from its circumference $|\mathbf{C}_h|$:

$$d = \frac{|C_h|}{\pi} = \frac{\sqrt{C_h.\ C_h}}{\pi} = \frac{a\sqrt{n^2 + nm + m^2}}{\pi} \tag{2.47}$$

In particular, the diameter cannot be regarded as a unique parameter for characterizing CNTs as different chiralities can generate the same nanotube diameter. The other two geometrical parameters (*T and θ*) can be derived from the chiral vector. For instance, the chiral angle is the angle between the chiral vector and the primitive lattice vector a_1:

$$\tan\theta = \frac{\sqrt{3}\, m}{(2n + m)} \tag{2.48}$$

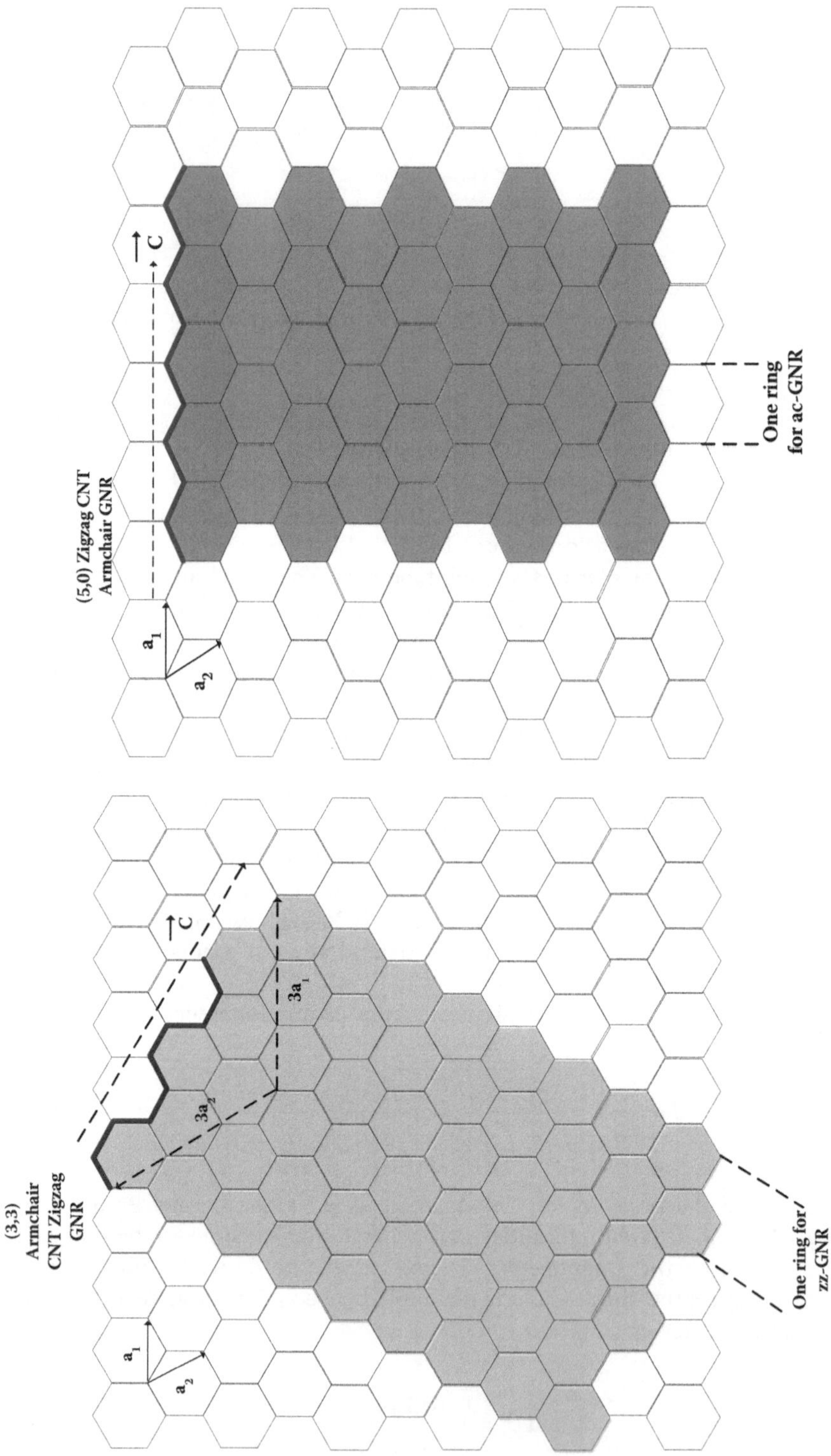

FIGURE 2.15 Types of CNT (armchair nanotubes (n=m), zigzag nanotubes (n=0 or m=0).

The translation vector defines the periodicity of the lattice along the tubular axis. Geometrically, **T** is the smallest graphene lattice vector perpendicular to C_h, and is expressed as $\mathbf{T} = t_1 a_1 + t_2 a_2$, where t_1 and t_2 are integers.

Thus, the chiral and translation vectors define the primitive unit cell of the CNT, which is a cylinder with diameter *d* and length *T*. Some auxiliary results that are useful to compute include the surface area of the CNT unit cell, the number of hexagons per unit cell and the number of carbon atoms per unit cell.

The summary of mathematical formulas representing the various ways of pivoting graphene into the tubes and defining structural parameters of CNT are expressed in Table 2.4 (Naeemi et al., 2005).

The Methods for CNT Synthesis

In past years, fullerenes were created by means of vaporizing graphite using a short-pulse, high-power laser method, while tubes of carbon were produced using carbon combustion and vapor deposition processes. By adopting the primitive techniques of creating fullerenes and CNTs in reasonable amounts, there is an application of an electric current across two carbonaceous electrodes in an inert gas atmosphere. The method of plasma arcing is fundamentally employed in order to create fullerenes and CNTs from various carbonaceous materials such as graphite. Here, the fullerenes appear in the soot, whereas CNTs are deposited on the electrode. In addition to this, a plasma arcing technique can also be applied in the presence of cobalt with 3% or greater concentration. There are various methods to synthesize CNTs, as depicted in the chart (Figure 2.16). With the application of these techniques, a variety of variable-diameter CNTs with different electronic and solid-state properties can be grown. Indeed, the synthesis of CNTs with a specific diameter or

TABLE 2.4
Structural parameters of CNTs

Symbol	Name	Formula	Value
a	Lattice constant	$a = \sqrt{3}a_{c-c} \approx 2.46$ Å	$a_{c-c} \approx 1.42$ Å
$\mathbf{a_1,a_2}$	Basis vectors	$(\sqrt{3}/2;1/2)a$, $(\sqrt{3}/2;-1/2)a$	_
$\mathbf{b_1,b_2}$	Reciprocal-lattice vectors	$\{(1/\sqrt{3};1)2\pi/a\}$, $\{(1/\sqrt{3};-1)2\pi/a\}$	_
$\mathbf{C_h}$	Chiral vector	$C_h= na1+ma2 \equiv (n,m)$	$(0\leq\lvert m\rvert \leq n$
d	Tube diameter	$d =\lvert C_h\rvert /\pi$	_
θ	Chiral angle	$\tan\theta = (\sqrt{3}m/(2n+m))$	$0\leq\lvert \theta\rvert \leq\pi/6$
T	Translational vector	$T=t_1a_1+t_2a_2\equiv(t_1,t_2)$	greatest common divisor $(t_1,t_2) =1$[a]
		$t_1=(2m+n)/N_R, t_2=-(2n+m)/N_R$	$N_R=gcd(2m+n,2n+m)$[a]
$\mathbf{N_c}$	Number of C atoms per unit cell	$Nc= \{4(n^2+nm+m^2)/N_R$	_

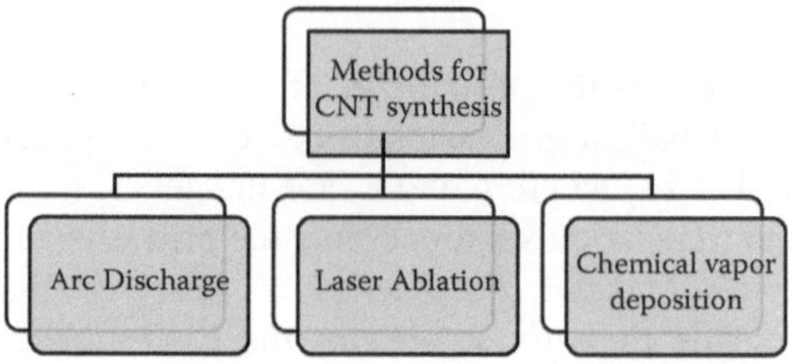

FIGURE 2.16 Methods for CNT synthesis.

chirality is perhaps the holy grail of nanotube science. This is a matter of further research with wide-ranging implications for both fundamental science and the accelerated development of industrial applications.

Out of these three methods, arc discharge and laser ablation are the well-known primitive techniques for CNT fabrication, whereas the chemical vapor deposition technique is a novel method for fabricating scale-up defect-free CNTs with promising electrical characteristics for integration into on-chip interconnect systems.

In this section, we will discuss in brief these three common techniques to yield nanotubes.

Arc Discharge Method Arc discharge, as shown in Figure 2.17(a), is the simplest

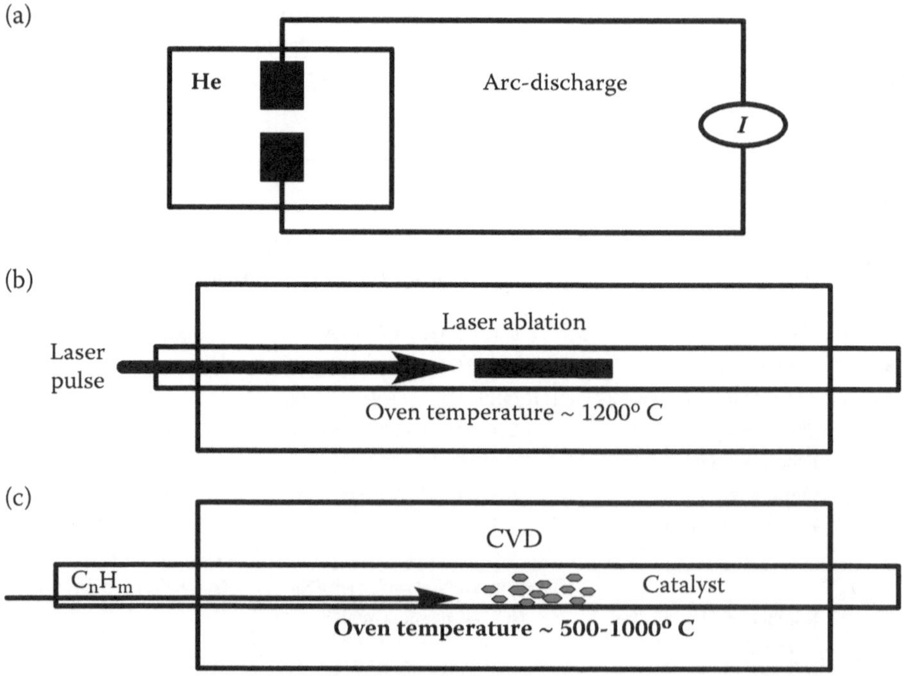

FIGURE 2.17 Techniques to yield CNT.

technique adopted for CNT growth. Here, CNTs are produced by keeping two carbon rods end to end in an enclosure and separated by a 1 mm distance. The inert gas is filled in the enclosure, at low pressure. By applying 50 to 100 amperes of direct current, a high temperature discharge is created between two electrodes. This high temperature discharge vaporizes the surface of the carbon electrodes and a small rod-shaped electrode is finally obtained. Lamentably, evaporation of carbon atoms in an arc discharge is a non-equilibrium process, and thus these atoms should undergo sequential condensation at temperatures below the melting point of graphite (3000 K). The production of CNTs in high yield primarily depends on the uniformity of the plasma arc. Carbon nanotubes arise as one of the products.

Laser Ablation Method In late 1990s, a dual-pulsed laser was employed in the synthesization method adopted for creating 70% pure CNTs. Nowadays, the laser vaporization technique is adopted for creating CNTs. The laser ablation method shown in Figure 2.17(b) deploys a high intensity laser pulse for ablating a targeted graphite rod consisting of 0.5 atomic percent of catalytic mixtures of cobalt and nickel. The carbon target rod is kept in a heated tube-furnace with a temperature of 1200°C. Flowing argon is passed through the chamber. The uniform vaporization of the target can be achieved by using the initial laser vaporization pulse followed by a second pulse. The amount of deposition of carbon as soot is primarily minimized by the usage of these two successive laser pulses. The larger particles are broken by applying the second laser pulse. The CNTs produced through this process are 10–20 nm in diameter and 100 μm or more in length. The average nanotube diameter and size distribution can vary for different growth temperature, catalyst composition and other process parameters.

Although both of the techniques discussed can produce high-quality CNT samples with a diameter raging from 2 to 200 nm and length ranging from 1 to 100 μm, CNTs grown by these methods suffer from certain drawbacks when applied to interconnect applications. It has been observed that CNTs obtained using these methods are lying horizontally on a flat substrate in a slow and random process. This restricts the tubes to be side-contacted electron transport across a Schottky Barrier formed at the metal–CNT junction. Therefore, a more robust and consistent architecture for CNT growth must be used for their applications in VLSI.

Chemical Vapor Deposition Method In the last two decades, production of carbon fibers and filaments are done by adopting the technique of chemical vapor deposition of hydrocarbons in the presence of a metal catalyst. Adoption of this technique leads to the production of a huge count of CNTs using catalytic CVD of acetylene over cobalt and iron. The bundles of singled-walled nanotubes of carbon along with the multi-walled nanotubes of carbon and fullerenes can be produced using the carbon/zeolite catalyst. Many studies have been reported for the formation of singled-walled nanotubes of carbon as well as multi-walled nanotubes of carbons from ethylene that is supported by the catalysts such as iron, cobalt and nickel. Furthermore, some recent research studies have also showcased the production of nanotubes on molybdenum and molybdenum–iron alloy catalysts. Figure 2.17(c) depicts the experimental setup for chemical-vapor-deposition growth. Growth is initiated after the introduction of gases into the CVD chamber, built up of high-temperature quartz material. Inside the high-temperature (~500–1000°C) sealed furnace, metal particles are used to catalyze the chemical vapor decomposition of a carbon-containing C_nH_m gas (such as methane, ethylene, acetylene, carbon monoxide, etc.) at atmospheric pressures. The metal catalyst is uniformly distributed. The carbon carrier gas nucleates on the surface of the metal particles, which catalyze the dissociation of the carbon gas with the freed carbon atoms arranged in a cylindrical manner to form a nanotube. Subsequently, after growth, the furnace is cooled to low temperatures (<300°C) before exposing the CNTs to air in order to prevent any oxidative damage to the nanotubes. The catalyst particle is usually found as carbide at the base or tip of the nanotube, depending on the adhesion of the particle to the substrate surface. In some cases, the particle can be somewhere around the middle of the nanotube, indicating that carbon atoms extend out in opposite directions of the catalytic particle. The diameter distribution of the nanotubes produced is often related to the statistical distribution of the particle sizes, and the length of the nanotube is dependent on the duration of growth. Growth rates can vary by many orders of magnitude, from several nanometers per minute to hundreds of micrometers per minute. Elucidation of the growth chemistry is a matter of ongoing research. The precipitation of carbon from the saturated metal particle leads to the formation of tubular carbon solids. Tube formation is favored over formation of other forms of carbon such as graphene sheets with open edges. This is because a tube contains no dangling bonds. A relatively high temperature is needed to form single-walled nanotubes with small diameters and allow the production of nearly defect-free nanotube structures. In the end, it can be concluded that the CVD process is very flexible and can be optimized to produce randomly oriented nanotubes as well as horizontally and vertically aligned nanotubes suitable for VLSI applications.

In addition to the three main techniques to synthesize CNTs, another commonly used methodology is ball milling. Ball milling is the uncomplicated technique to manufacture CNTs followed by subsequent annealing. Using carbon and boron nitride powder, CNTs are produced by thermal annealing. The procedure adopted in this method is that firstly, in a stainless steel container, powder of graphite is placed. This stainless steel container consists of four hardened steel balls. The steel container is purged and at ambient temperature argon is introduced for milling process for up to 150 h. Thus, under an inert gas flow for 6 hours at 1400°C, the annealing of graphite powder is done. This method is suitable for synthesizing more multi-walled nanotubes and very few single-walled nanotubes. Other methods for producing CNTs are electrolysis, synthesis from bulk polymer, flame synthesis, use of solar energy and low-temperature solid pyrolysis. In the electrolysis technique, by passing an electric current in molten ionic salt between graphite electrodes, production of CNTs are carried out. A high purity carbon rod serves as a cathode. The cathode gets consumed and a wide range of nonmaterial is produced. Synthesis of CNTs can also be carried out chemically by using polymers consisting of carbon. This technique results in production of multi-walled nanotubes of diameters from 5 nm to 20 nm and a length up to 1 micrometer. In the flame synthesis technique, by providing a portion of hydrocarbon gas at an elevated temperature, the hydrocarbon reagents function. The methodology adopted in synthesizing CNTs using the pyrolysis method is to obtain nanotubes of carbon from metastable carbon containing compounds at relatively low temperatures of 1200°C–1900°C. From this method, multi-walled nanotubes with diameters ranging from 10 nm to 25 nm and lengths of 0.1–1 micrometer are obtained.

The Methods for CNT Purification

Ensuing to the synthesis process, the purification of CNTs are carried out by segregating them from other entities like carbon nanoparticles, residual catalyst, amorphous carbon and other undesired species. The traditional chemical methodology of purification seems to be ineffective for the purpose of eradicating the unwanted impurities from CNTs. Henceforth, three fundamental techniques known as the gas phase, liquid phase and intercalation, are primarily adopted in order to purify the CNTs from unwanted impurities. To be more generalized, an operation called microfiltration is carried out in order to eradicate the nanoparticles with the use of membrane filters. This technique, without chemically changing the CNTs, removes simultaneously the carbon amorphous and undesired nanoparticles. Moreover, the minute chemical substances can be removed with the application of 2–3 mol. nitric acid. By the use of extended sonication within the concentrated acid mixtures, nanotubes of carbon are sliced into minute segments, forming a colloidal suspension in solvents. The solvents can be deposited on substrates or in a solution that provides different functional groups attached to the sides and ends of CNT. A brief description of a chiefly used purification method is discussed in the next paragraph.

Gas Phase Purification Technique The gas phase technique was invented by Thomas Ebbesen, along with his co-workers. It was known to be the most successful methodology adopted for purification. By adopting this technique, the researchers came to the conclusion that the oxidization process is quite easier for the nanoparticles with defects in comparison to the relatively perfect nanotubes. This method assists in producing a significant improvement of carbon nanotubes.

In the recent past, NASA GRC has come up with a novel gas phase technique for purification of single-walled nanotubes in gram-scale quantities. This methodology works in a combination of high-temperature oxidation and repeated extraction with nitric and hydrochloric acid. By adopting this improved methodology, the stability of CNTs are significantly enhanced with a minute reduction of impurities like residual catalyst and non-nanotube forms of carbon.

Liquid Phase Purification Technique The second most effective technique for purification is the liquid phase purification method. The step-by-step procedure followed in the liquid phase purification technique is as mentioned below:

1. Preliminary filtration: It is the initial step taken to remove graphite particles that are very large.
2. Dissolution: After filtration, the next step will be to remove the fullerenes and catalyst particles.

3. Microfiltration: Thirdly, this step taken is to perform microfiltration.
4. Centrifugal separation: The next step is to carry out separation by exploiting the physics of the centrifugal phenomenon.
5. Chromatography: Finally, the chromatography is done so as to achieve pure CNTs after liquid phase purification.

Thus, in this method, CNTs go through liquid phase oxidation in the presence of a hydrogen peroxide solution. Moreover, as it removes the amorphous carbon without damaging the tube walls, hence it is easier to separate CNTs in the final stage of separation.

Intercalation Purification Technique Harris, along with his research team, introduced this method in the late 1990s. In this method, a molecule (or ion) is inserted into compounds with layered structures. The outcome of this technique is the expansion of van der Waals gap in between adjoining layers, which requires energy that is supplied by transfer of charge between the solid and dopants.

By an intercalation with copper chloride, this makes the oxidization of nanoparticles easier. Henceforth, this makes it quite a popular technique for the purification of nanotubes. The procedure adopted is to first immerse the cathode in a mixture of molten copper chloride with potassium chloride for about a week at a temperature of 400°C. For removing the excess of copper chloride and potassium chloride, ion-exchanged water is used for washing the mixture of graphitic fragments and the intercalated nanoparticles. Furthermore, to mitigate the intercalated copper chloride and potassium chloride metal, the washed product is briskly heated at a temperature of 500°C for an hour in a mixture of helium and hydrogen. Lastly, oxidation of material is done in a flowing air with a rate of 10°C/min and at a temperature of about 555°C. This technique produces a fresh sample of cathodic soot that is taken as a fresh nanotube. The major shortcomings of this technique are it results in a loss of some amount of nanotubes at the oxidation stage and secondly it leads to contamination of the final material with residues of intercalates.

The Attributes of CNTs

The fashion in which the carbon–carbon (C–C) atoms are arranged in nanotubes directs and determines the unique electrical and mechanical as well as thermal properties of CNTs. These attributes are briefly discussed below.

a Electrical Resistivity As aforementioned, the resistivity of CNT structures is a function of its chirality. Thus, degree of twist of a graphene sheet defines the resistivity of CNT interconnects. Usually metallic CNT are considered highly conductive. The carbon nanotubes can exhibit both semiconducting as well as metallic properties based on chiral indices.

b Strength and Elasticity In a single sheet of graphene, the carbon lattice has a carbon-carbon atom connected through a strong chemical bond to the three adjacent carbon atoms. In this manner, a nanotube of carbon evinces the strongest basal plane elastic modulus, thereby expected to be an ultimate high strength fiber. The elastic modulus of single-walled carbon nanotubes is much higher than steel, which makes them highly resistant. Although pressing on the tip of a nanotube will cause it to bend, the nanotube returns to its original state as soon as the force is removed. This property makes CNTs extremely useful as probe tips for high-resolution scanning probe microscopy and also for flexible interconnection.

c Thermal Resistivity and Expansion The nanotubes of carbon usually demonstrate superconductivity below 20K because of the strong in-plane carbon-carbon bonds present in a sheet of graphene. These strong carbon-carbon bonds will result in an exceptional strength and stiffness against axial strains. Apart from this, the greater interplane and zero in-plane thermal expansion of single-walled carbon nanotubes leads to high flexibility against non-axial strains. Due to their high thermal conductivity and large in-plane expansion, CNTs exhibit exciting prospects in nanoscale molecular electronics, sensing and actuating devices, reinforcing additive fibers in functional composite materials, etc. Therefore, it is expected that the nanotube may also significantly improve the thermo-mechanical and the thermal properties of the composite materials.

d Aspect Ratio One of the most promising attributes of nanotubes made up of carbon is the high value of aspect ratio. This means that a lower CNT load is required compared to other conductive additives to achieve similar electrical conductivity. The high aspect ratio of CNTs possesses unique electrical conductivity in comparison to the conventional additive materials such as chopped carbon fiber, carbon black or stainless steel fiber.

e Absorbent Carbon nanotubes and CNT composites have been emerging as perspective absorbing materials due to their light weight, larger flexibility, high mechanical strength and superior electrical properties. Therefore, CNTs emerge as ideal candidate for use in gas, air and water filtration.

2.4.2.2 Dominion of GNR Interconnect

Apart from carbon Nanotubes (CNT), even graphene nanoribbons (GNR) are also ruminated by various researchers as an embryonic alternative for patterning on-chip interconnect structures (Yagmurcukardes et al., 2016). GNRs are basically very fine strips of thin-width graphene layers and can be visualized as mutant forms of unrolled CNTs. To a great extent, the GNRs exhibit similar electrical and physical traits as that of CNTs; however, the astonishing characteristics of GNRs which make it distinguishable to the latter is due to its planar geometry both interconnecting wire structures as well as transistors can be implemented on the very same graphene layer (Figure 2.18) (Banadaki, 2016). Hence, it circumvents the manufacturing abnormalities related to metal-nanotube contact. As aforementioned, the GNRs will outperform the Cu interconnects for smaller widths due to its commendable electrical properties, larger mean free path (1×10^3 nm) and better current-carrying capacity ($5–20 \times 10^8 A/m^2$).

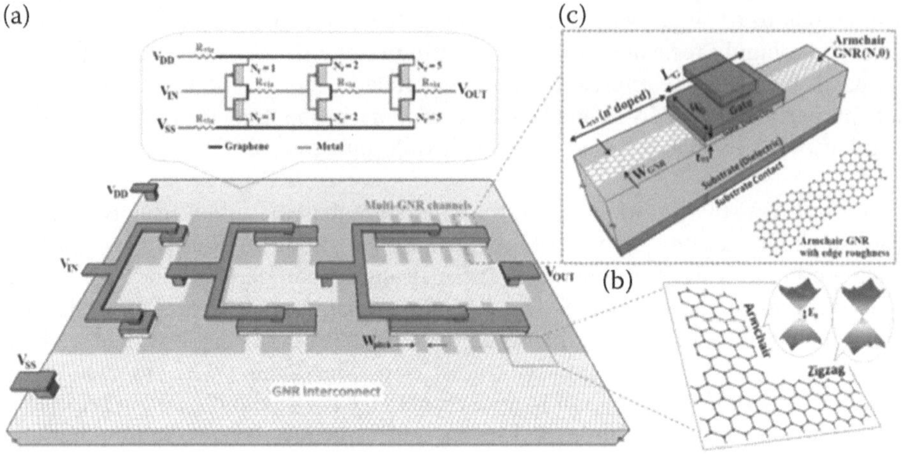

FIGURE 2.18 3D view of all-graphene circuit: (a) an example of inverter chains together with its circuit implementation. The contact resistances of via connections are also shown corresponding to the layout design. (b) Narrow armchair-edge GNRs and wide zigzag-edge GNRs are used as channel material and local interconnects, respectively. (c) 3D view of a GNRFET with one ribbon of armchair GNR (N, 0) as channel material.

Basic Structure of GNRs

Graphene is an allotrope of carbon from which thin layers of ribbons can be obtained. These narrow-width conducting ribbons of an ultra thin (<50 nm) graphene layer are termed graphene nanoribbons (GNRs). A graphene nanoribbon is a single sheet of graphene layer, which is extremely thin and limited in width, such that it results in a one-dimensional structure (Nakada et al., 1996). Mitsutaka Fujita, in 1996, along with his group, has provided a pedagogical model of graphene nanoribbons (GNRs) to study the effect of nanoscale dimensioning in graphene. Depending on termination of their width, GNRs can be divided into chiral and nonchiral GNRs.

Secondly, unlike the difficulty encountered with CNTs, the chirality control is easier in the case of GNR interconnects. On the basis of chirality, GNRs are also classified as zigzag (zz) or armchair (ac), but this classification is antithesis of terminology used for CNTs, such that nomenclature “zigzag” or “armchair” used in the case of the former shows the termination style, i.e. structure of GNR edge, whereas for the latter the same terminology indicates the circumference of the tube. In other words, a zigzag GNR can be interpreted as an unrolled version of an armchair CNT and vice versa, as depicted in Figure 2.19.

The width of GNR is determined by the number of dimer lines (N). To quantify the magnetic, electric and optical properties on graphene nanoribbons, various theoretical models have been applied. In context to the tight binding model theory, it is predicted that zigzag nanoribbons of graphene are inevitably metallic while the conductivity of armchair-type GNRs is determined by the total count (N) of hexagonal carbon rings along the width (Xu et al., 2008). Figure 2.20 shows the

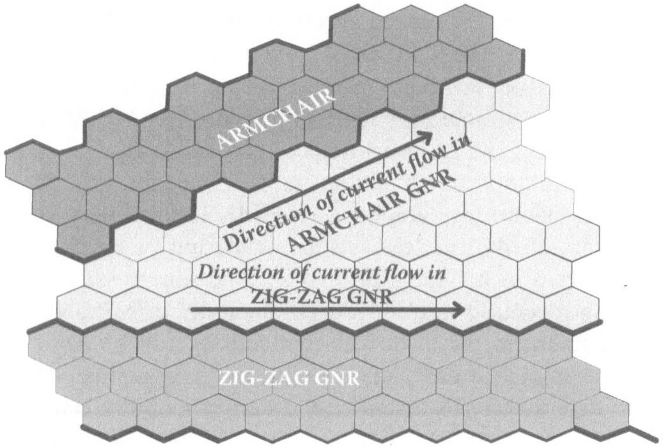

FIGURE 2.19 GNR structure and its chirality.

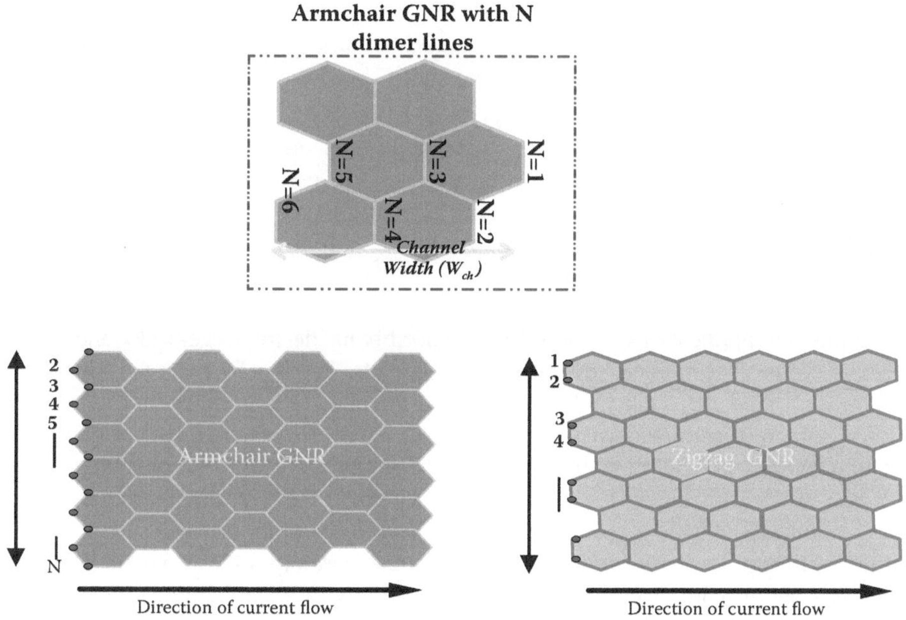

FIGURE 2.20 Types of GNRs.

structure of ac-GNR with the number of carbon atoms across its width, N = 6. If N = 3p–1 or 3p+2, then ac-GNR exhibits metallic behavior, whereas it acts as a semiconductor if N= 3p or 3p+1 (p is an integer). Predominantly semiconducting GNR has a 0.2eV bandgap, while metallic ones have a zero bandgap.

Subjected to the number of stacked sheets of graphene, GNRs can be grouped into two types: single-layer GNR (SLGNR) and multi-layer GNR (MLGNR). The monolayer-type SLGNR offers high resistivity; hence, it is not suitable for designing interconnect wire structures, whereas the multilayer MLGNRs are preferred as they are easier to fabricate and also reduce the overall equivalent resistance of wire by providing multiple conduction paths (Naeemi and Meindl, 2007). Nonetheless, of this patterning, graphene into MLGNR renders edge roughness (Nishad and Sharma, 2014) that leads to the scattering of electrons and thereby diminishes the effective mean free path (MFP). The repercussion of reduction in MFP is a significant decrease in electrical conductivity of MLGNRs interconnect wires. The enhancement in the electrical conductivity can be achieved either by enriching the carrier mobility or by augmenting the total number of carriers. It is being reported to date that a substantial rise in carrier mobility can be obtained by intercalation doping of pristine MLGNRs with AsF5 or Li (Nishad and Sharma, 2015). The number of carriers rises by using the above technique because it causes an injection of charges where carriers have less mobility from the intercalated layer to the MLGNR layer where there is high mobility. As an effect of this, the Fermi level shifts upward and interlayer spacing increases, resulting in low scattering and higher MFP.

Moreover, with regards to their coupling with adjacent devices, MLGNR can further be categorized as a top contact GNR and side contact GNR (Hamedani and Moaiyeri, 2020). Side contact MLGNRs show better performance as all the stacked graphene layers are joined physically to contact, thus offering low resistance and high conduction channels. While in the case of top contact MLGNRs, only the uppermost layer is connected with the contacts and hence equivalent resistance is high, which degrades performance (Hamedani and Moaiyeri, 2019) (Figure 2.21).

The Methods for GNR Synthesis

The industrial applications of graphene nanoribbons desire large-scale and cost-effective synthesis techniques, eventually catering a trade-off in between the simplicity of fabrication and the quality of final material obtained. It is defined as on-demand tailored properties as in accordance to the ultimate use. The chief benefit of graphene is that in contrast to other nano materials, graphene can be synthesized on a huge scale with cost-efficient methods. In order to synthesize, both the bottom-up approach (like atom by atom growth) or top-down approach like exfoliation from bulk are employed. Before going up for mass production, generating laboratory-scale protocols for graphene production with targeted properties is necessary.

There are various methods to synthesize atomic-layer graphene-like mechanical exfoliation of graphite, deposition of epitaxial graphene on SiC crystals and chemical vapor deposition of graphene using a metal catalyzer. Out of these three methods, the simplest method is mechanical exfoliation, which is also known as micromechanical cleavage (MC). In this method, graphene layers are peeled repeatedly using the popular scotch-tape technique (Novoselov et al., 2004) to achieve a graphene monolayer that could then be transferred to an oxidized substrate. MC can be optimized in order to yield good-quality layers, with size restricted by the single crystal grains in the starting graphite, in the range of millimeters. The number of ribbons is

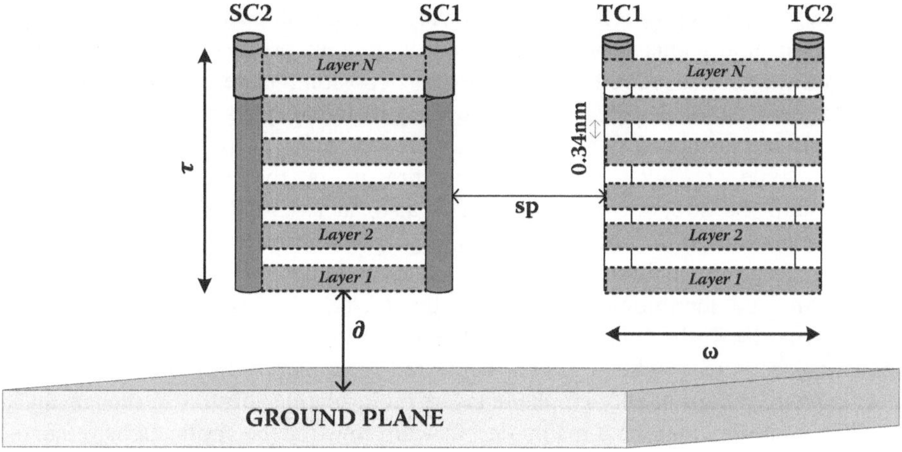

FIGURE 2.21 SC and TC MLGNR interconnect model.

identified by elastic that is contrast spectroscopy and inelastic light scattering that is Raman spectroscopy. Although the mechanical exfoliation is the simplest technique, unfortunately it cannot be used for mass production. Indeed, the large majority of basic prototype applications are generated with MC flakes. Another commonly used fabrication method is direct deposition of epitaxial graphene on insulating SiC crystals substrate by sublimating silicon atoms at high temperatures and thus preventing the need for the transfer process (Emtsev et al., 2009). The technique of graphite synthesis from SiC was initially carried out in the early 1890s. Growth of graphene on SiC is termed epitaxial growth, despite the fact that there is a greater amount of mismatch between SiC (3.073Å) and graphene (2.46 Å) and, moreover, the carbon rearranges itself in a hexagonal framework once the Si starts evaporating from the SiC substrate, rather than being deposited on the SiC surface as in the case of conventional epitaxial growth process. The weakness of this technique is that the cost rises exponentially for extensive production in the case of SiC as compared to the same size Si wafers. The strategy to lessen the substrate cost is to grow thin SiC layers on sapphire. Other points that need to be addressed are plans to improve growth of cubic SiC as a substrate, and the growth of insulating SiC layers on cheap on-axis *N*-type substrates, in order to replace expensive semi-insulating substrate materials. The benefit over standard CVD is the graphene quality control achievable by the tuning of carbon source thickness and annealing conditions. In addition, all of the process steps occur in a fully semiconductor compatible environment. Thus, industry can then benefit from the versatility of this method to integrate graphene in their process flow. The long-term goal is a totally controlled graphene nanostructuring, so as to produce GNRs on demand. This is motivated by the prospect of bandgap creation in graphene.

In addition to the above-mentioned techniques, carbon can also be deposited on the metallic surface with the help of numerous methods like spin coating, chemical vapor deposition, flash evaporation (PVD) and chemical vapor deposition (CVD).

Chemical vapor deposition is a process widely used to deposit or grow thin films, crystalline or amorphous, from solid, liquid or gaseous precursors of many materials. CVD is said to be the workhorse for depositing materials used in semiconductor devices for decades. There are many different types of CVD processes: thermal, plasma enhanced (PECVD), cold wall, hot wall, reactive and many more. Again, the type depends on the available precursors, the material quality, the thickness and the structure needed; cost is also an essential part of selecting a specific process. In 1984, the first studies of graphene growth on metals via CVD was undertaken. The main objective of this study is to develop an understanding of the catalytic and thermionic activities of the metal surfaces in the presence of carbon. After 2004, the focus shifted to the actual growth of graphene. Electronic applications require graphene grown, deposited or transferred onto dielectric surfaces. Growth of high-quality graphene layers on insulating substrates, such as SiO_2 and SiC sapphire, would be a major step forward towards the applications in nanoelectronics. As mentioned previously, growth of a single crystal graphene on dielectric surfaces is highly desirable but to date the crystal size on dielectrics is limited to micron size. Plasma-enhanced CVD (PECVD) is a scalable and cost-effective, large-area deposition technique, with numerous applications ranging from electronics (IC, interconnects, memory and data storage devices), to flexible electronics and photovoltaic.

Thus, the CVD method is preferred as it leads to growth of large-area graphene which is in turn favorable for wafer-scaled lithography (Obraztsov, 2009). In the CVD process, hydrocarbon molecules supply the carbon atoms that are dissolved on the metal surface and then transferred to an insulating wafer. Using e-beam lithography, the fabrication of a narrow strip of graphene can be done and then the width of a graphene nanoribbon can be further reduced by etching down to 4 nm and 2 nm by chemical synthesis (Han et al., 2007) with a very smooth edge. Other lithography methods based on atomic force microscopy and scanning tunneling microscopy (STM) have been proposed for the fabrication of GNRs. GNRs of 1 nm width can be produced by unzipping carbon nanotubes with a bottom-up chemical approach (Xie et al., 2011). In principle, graphene can be chemically synthesized, assembling benzene building blocks. In such an approach, small organic molecules are linked through surface-mediated reactions at a relatively low temperature (<200°C). The resulting materials include nanostructured graphenes, which may be porous, and may also be viewed as 2D polymers. The chemical approach offers opportunities to control the nano-graphenes with well-defined molecular size and shape, thus, properties that can be tuned to match the requirements for a variety of applications, ranging from digital and RF transistors, photo detectors, solar cells, sensors, etc. GNRs with a well-defined bandgap with tuneable absorption can already be designed and produced. Such approaches will ultimately allow control at the atomic level, while still retaining the essential scalability to large areas. Other challenges for direct chemical growth include growth on insulating substrates for electronic applications and development for efficient transfer methods that allow the nanoribbons to be incorporated in electronics. GNRs can be obtained by combining e-beam lithography and oxygen plasma etching. Graphene nanoribbons of size reduced down to ~20 nm were reported, with a bandgapgap of ~30 meV, are then used in FETs with I_{ON}/I_{OFF}

up to 103 at low T (<5K) and ~10 at room temperature. Chemical synthesis seems to be the most promising route towards well-defined GNRs.

2.5 ELECTRICAL IMPEDANCE MODELING OF ON-CHIP INTERCONNECTS

The impedance characteristics of an on-chip interconnect includes the resistance, capacitance and inductance. For an accurate estimation of these wire parasitics, a detailed study of interconnect geometry and material property is necessary. Much of the discussion related to material property has already been done in preceding sections. This section describes the parasitic formulation of copper, CNT (carbon nanotube) and GNR (graphene nanoribbon) interconnect impedance models.

2.5.1 Geometry of Conventional Copper Interconnect and Impedance Calculation

Figure 2.22 shows the pair of adjacent copper interconnect wires. The interconnect wires have length l, thickness t, width w, the distance between center of signal line and ground shield dsg and the distance between two ground shields as dgg. The sum of the width and spacing is called the *wire pitch*. The thickness to width ratio, t/w, is called the *aspect ratio*. The parameters and interconnect dimensions of Cu are mentioned in Table 2.5.

The analytical expression to calculate equivalent impedance parameters of rectangular copper interconnects are formulated in the remainder of this section.

2.5.1.1 Interconnect Resistance Estimation

The equivalent resistance of a uniform rectangular slab of copper interconnects having area A and resistivity ρ is expressed as

$$\mathbf{R} = \frac{\rho . l}{A} = \frac{\rho . l}{tw} = R_{\square}\frac{l}{w} \tag{2.49}$$

where $R_{\square}$ is the sheet resistance and has units of $\Omega/square$. Here, a square is a dimensionless quantity corresponding to a slab of equal length and width. This is convenient because resistivity and thickness are characteristics of the process

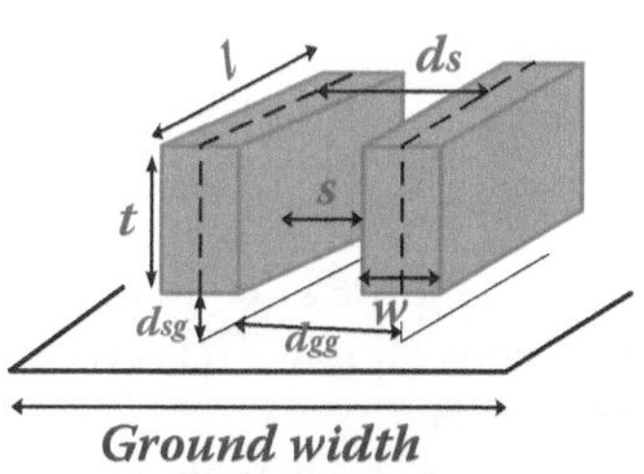

FIGURE 2.22 Copper interconnect geometry.

TABLE 2.5
Interconnect dimensions and structural parameters of Cu interconnects at 32 nm technology node

LOCAL	Width (nm)	32
	Aspect Ratio (A/R)	1.8
	Thickness (nm)	57.6
	Height (nm)	54.4
	ρ_{cu} (μΩ.cm)	2.2
	Dielectric constant (ε_r)	2.2
GLOBAL	Width (nm)	48
	Aspect Ratio (A/R)	2.34
	Thickness (nm)	112.32
	Height (nm)	72
	ρ_{cu} (μΩ.cm)	2.2
	Dielectric constant (ε_r)	2.2

outside the control of the circuit designer and can be abstracted away into the single-sheet resistance parameter.

Due to specialized processing and operating conditions of the on-chip copper interconnect, certain non-ideal effects need to be considered, making the effective resistivity deviate from the ideal bulk resistivity.

Diffusion Barrier

As shown in Figure 2.23, a thin and highly resistive barrier layer is built on three sides of an on-chip Cu interconnect to prevent Cu from diffusing into the surrounding dielectric. This barrier layer consumes part of the cross-sectional area allocated to the interconnect, thereby reducing the effective wire cross-sectional area and hence raises the resistance. Moreover, *chemical mechanical polishing* is done to smooth and planarize the surface and further causes *dishing* that thins the metal. If the average barrier thickness is $t_{barrier}$ and the height is reduced by t_{dish}, the resistance becomes

$$R = \frac{\rho}{(t - t_{dish} - t_{barrier})} \cdot \frac{l}{(w - 2\,t_{barrier})} \tag{2.50}$$

Surface and Grain Boundary Scattering

When the dimensions of the interconnect are scaled deep into the DSM regime, the resistivity of the interconnect increases as the wire dimensions shrink. The reason behind this is due to the phenomenon illustrated in Figure 2.24.

As the wire cross-section dimensions and grain size become comparable to the bulk MFP of electrons in Cu, two major physical phenomena (namely surface

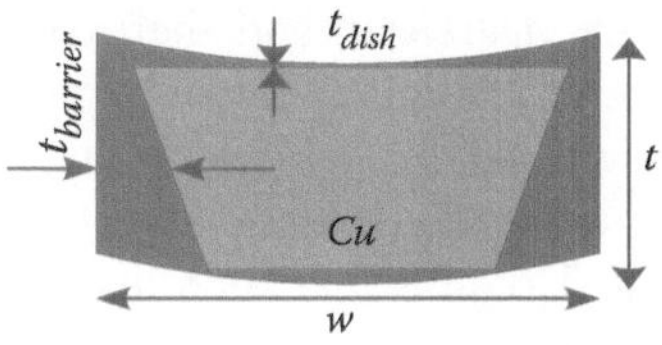

FIGURE 2.23 Cross section of copper interconnect showing copper barrier layer and dishing.

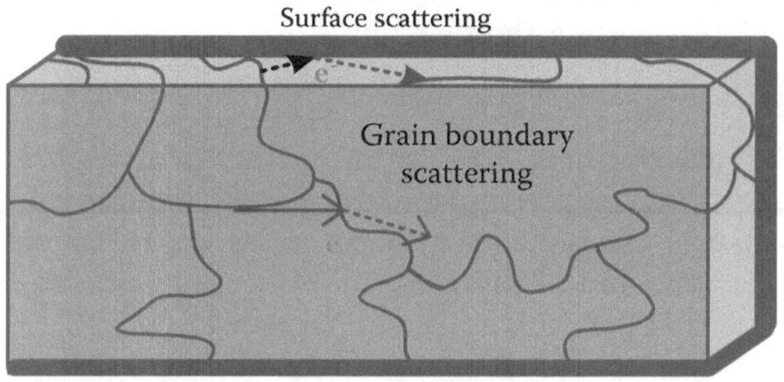

FIGURE 2.24 Schematic illustrations of the surface and grain boundary scatterings.

scattering and grain boundary scattering) occur to the current-carrying electrons in Cu in addition to bulk phonon. The Fuchas–Sondheimer model and the theory of Mayadas and Shatzkes quantify these two effects. Thus, considering the effect of surface roughness and grain boundary scattering phenomenon, the resistivity can be formulated using the Fuchas–Sondheimer surface roughness resistivity model (ρ_{FS}) (Sondheimer, 1952) and the Mayadas and Shatzkes grain boundary scattering resistivity model (Mayadas and Shatzkes, 1970) (ρ_{MS}) as

$$\rho_{Cu} = \rho_{FS} + \rho_{MS} \tag{2.51}$$

$$\frac{\rho_{FS}}{\rho_{bulk}} = 1 + 0.75\left[\frac{\lambda_{Con}(1 - P)}{w}\right] \tag{2.52}$$

$$\frac{\rho_{MS}}{\rho_{bulk}} = \left[1 - 1.5\Delta + 3\Delta^2 - 3\Delta^3 \ln(1 + \frac{(1)}{\Delta})\right]^{-1} \tag{2.53}$$

$$\Delta = \frac{\lambda_{Con}}{D}\frac{R}{(1 - P)} \tag{2.54}$$

where bulk material resistivity is ρ_{bulk}, MFP is λ_{Con}, P is Fuschas scattering parameter, D is the size of mean grain and R is the reflection coefficient having values between 0 and 1.

Yet another significant problem that comes into the picture due to scaling of wire dimensions is susceptibility to electromigration. To delve into detail of this problem, it can be explained as the wire dimensions are decreased and then the cross-sectional area is also reduced; furthermore, because of a significant rise in the count of devices integrated on a chip, the chances of simultaneous switching also increases. As an outcome of all of this, current density of interconnects also rise considerably. With the increase in value of current density to an extent such that it crosses the threshold limit, the electrons knocked out some of the host metal atoms to migrate from their original location. This phenomenon is called *electromigration*. As a result of this electromigration, the thin metal film breaks, thereby creating voids, or form hillocks creating short circuits.

Temperature Effect

Studies show that the electrical behavior of circuits is affected by temperature fluctuations, mainly due to the effect of electron–phonon scattering on the mean free path (MFP), which determines the interconnect resistance. Henceforth, considering the effect of temperature on resistivity

$$\rho_{Cu}(T) = \rho_{Cu}(0) + \left(\frac{4\,R(\Theta_R)\,T^n}{\Theta_R^n}\right)\int_0^{\frac{\Theta_R}{T}} \frac{Z^p}{(e^Z - 1)(1 - e^{-Z})}dZ \tag{2.55}$$

$$R(\Theta_R) = \frac{h^3}{e^2}\left[\frac{\pi^3(3\pi^2)^{1/3}}{4n^{2/3}a\,M\,k_B\,\Theta_R}\right] \tag{2.56}$$

where the resistivity of nano Cu interconnects is calculated with Debye temperature Θ_R (320 K), h is Planck's constant, n is the number of electrons available for current conduction in an atom, M is atomic mass, k_B is Boltzmann's constant and e is the electron charge. p is an integer whose value depends on the characteristics of the interaction. Thus, effective resistance of a Cu interconnect after taking into consideration all the effects is given by

$$R_{Cu}(T, w_{int}) = \left(\rho_{Cu} + \rho_{Cu}(T)\right)\frac{l}{w \times t} \tag{2.57}$$

2.5.1.2 Interconnect Capacitance Estimation

For an ASIC on-chip interconnect, estimation of parasitic capacitance is a most trivial process. As shown in Figure 2.25, each interconnect is a 3D structure with significant variations of shape, thickness and vertical distance from the ground plane or substrate. Also, each interconnecting wire line is typically surrounded by a number of other lines, either on the same level or on different levels.

The accurate estimation of the parasitic capacitances of these interconnecting wires with respect to the ground plane or substrate, as well as with respect to each other, is quite tedious. Although 3D field solver, such as FastCap, provides an accurate capacitance result, it has the limitations of timing and memory overhead. Moreover, with a rapid rise in integration, the number and geometric complexity of

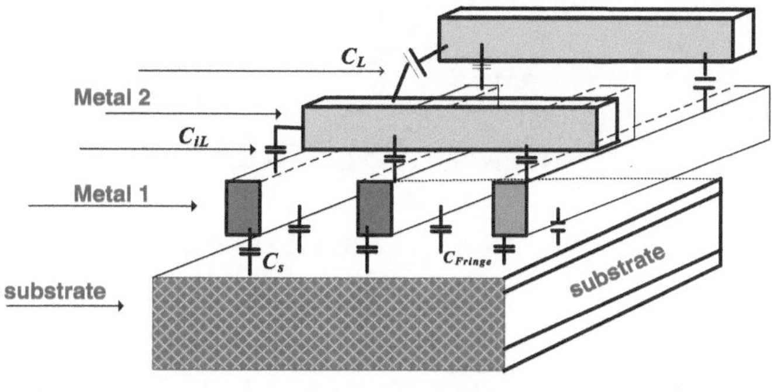

FIGURE 2.25 Simplified views of six interconnections on three different levels, running in close proximity to each other.

the on-chip interconnects drastically increases. It is, therefore, not practical to apply a field solver to an entire IC.

The interconnecting wire capacitance has two major components: the parallel plate capacitance of the bottom of the wire to ground and the fringing capacitance arising from fringing fields along the edge of a conductor with finite thickness. In addition, a wire adjacent to a second wire on the same layer can exhibit capacitance to that neighbor. To understand these components at first, consider the section of a single interconnect, which is shown in Figure 2.26. It is assumed that an interconnect segment of length l, width w and thickness t runs parallel to the surface and is separated from the ground plane by a dielectric layer of height h.

Now, the correct estimation of the parasitic capacitance with respect to ground is an important issue. Using the basic geometry illustrated in Figure 2.27, one can calculate the parallel-plate capacitance of the interconnect segment from the classic parallel plate capacitance formula as

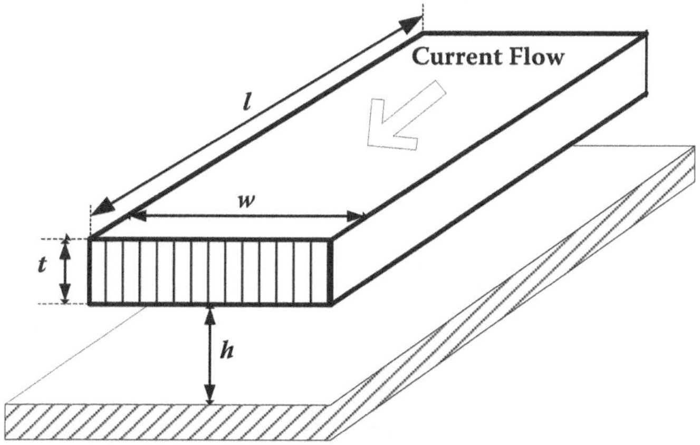

FIGURE 2.26 Interconnect segment running parallel to the surface, which is used for parasitic resistance and capacitance estimations.

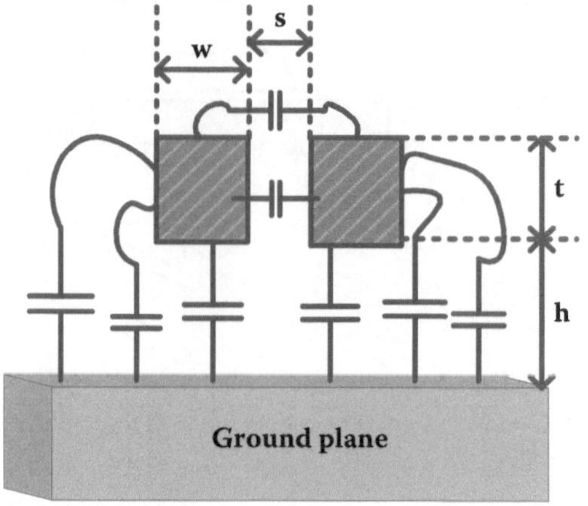

FIGURE 2.27 Effect of fringing fields on capacitance.

$$C = \frac{\varepsilon_{ox}}{h} w.\, l \tag{2.58}$$

However, in interconnect wire lines where the thickness of wire (t) is comparable in magnitude to the ground-plane distance (h), fringing electric fields significantly increase the total parasitic capacitance (Figure 2.27). It is a widely known fact that DSM technologies allow the width of the metal lines to be decreased rather significantly, but the thickness of the line must be preserved in order to ensure structural integrity. This situation, which involves narrow metal lines with a considerable vertical thickness, makes these interconnection lines especially vulnerable to fringing field effects.

A number of authors have proposed approximations to compute the fringing field capacitance calculation [Barke88, Ruehli73, Yuan82] (Ruehli and Brennan, 1973; Yuan and Trick, 1982). A set of simple formulas developed by Yuan and Trick in the early 1980s can be used to estimate the capacitance of the interconnect structures in which fringing fields complicate the parasitic capacitance calculation. Here, as depicted in Figure 2.28, a lone conductor above a ground plane is assumed to be a rectangular middle section with two hemispherical end caps.

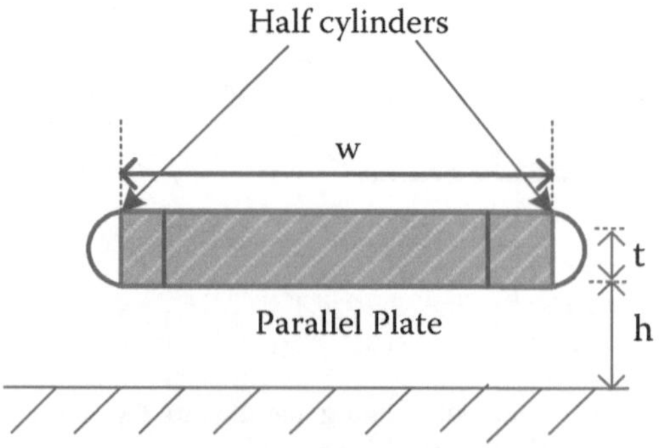

FIGURE 2.28 Yuan and Trick capacitance model including fringing fields.

The total capacitance is assumed to be the sum of a parallel plate capacitor of width $\left(w - \frac{t}{2}\right)$ and a cylindrical capacitor of radius $\left(\frac{t}{2}\right)$. This gives an expression for the capacitance that is accurate within 10% for aspect ratios less than 2 and $t \approx h$.

$$C = \varepsilon_{ox} l \left[\frac{w - \frac{t}{2}}{h} + \frac{2\pi}{ln\left\{1 + \frac{2h}{t} + \sqrt{\left(\frac{2h}{t}\left(\frac{2h}{t} + 2\right)\right)}\right\}} \right] \tag{2.59}$$

An empirical formula that is computationally efficient and relatively accurate is (Barke, 1988)

$$C = \varepsilon_{ox} l \left[\frac{w}{h} + 0.77 + 1.06\left(\frac{w}{h}\right)^{0.25} + 1.06\left(\frac{t}{h}\right)^{0.5} \right] \tag{2.60}$$

The above formula holds if it is assumed that the interconnection line is completely isolated from the surrounding structures. But in a real scenario for multi-level metal processes in an IC, the interconnect line is usually coupled with other lines running in parallel. In such cases, the total parasitic capacitance of the line is not only increased by the fringing-field effects, but also by the capacitive coupling between the adjacent, parallel neighboring lines. Due to scaling, when the thickness of the interconnect wire becomes comparable to its width, then the capacitive coupling between neighboring interconnect lines becomes prominent. This coupling between the adjacent interconnect lines is mainly responsible for signal crosstalk, when transitions in one line can cause noise in the other neighboring lines (detailed analysis of which is discussed in Chapter 4 of this book).

2.5.1.3 Interconnect Inductance Estimation

In most of the scenarios, only interconnect resistance and capacitance are considered for modeling on-chip impedance to calculate the Elmore delay, etc. As compared with resistance and capacitance, the interconnect inductance is significantly more difficult to extract and model, so engineers prefer to design in such a way that inductive effects are negligible. Nevertheless, inductance needs to be considered in high-speed designs for wide wires such as clocks and power buses (Kaustav and Mehrotra, 2002).

Generally, to extract inductance it is a three-dimensional problem and is quite time consuming, specifically for complex wire geometries. One reason for this difficulty is due to the loop-based inductance definition:

$$L_{ij} = \frac{\psi_{ij}}{I_j} \tag{2.61}$$

where ψ_{ij} is the magnetic flux in loop i induced by the current I_j in loop j. For the formation of a loop, identification of the current return paths has to be done. The current distribution in a circuit mainly depends on the interconnect characteristics. For the case of wide global interconnects, the inductance effect is quite significant in top metal layers rather than that of local interconnects in lower metal layers. The difficulty encountered in inductance extraction is mainly due to two reasons. First, the wires in adjacent layers are generally orthogonal hence the adjacent layers can no longer be treated as a ground plane as in capacitance extraction. Second reason is due to long-range inductive coupling effects. Artificially restricting the inductance extraction to nearby geometries not only introduces inaccuracy but may also result in unstable models. Thus, the pattern matching method used for capacitance extraction cannot be applied for inductance extraction due to the complex geometries surrounding the wire.

Hence, it can be stated that as inductance value depends on the entire loop and cannot be simply decomposed into sections as in case of capacitance, therefore it is impractical to extract the inductance from a chip layout. Instead, usually inductance is extracted using tools such as Fast Henry (Kamon et al., 1994) for simple test structures intended to capture the worst cases on the chip.

But, pedagogically, the analytical calculation for inductance of a conductor (Figure 2.22) of length l and width w located at height d_{sg} above a ground plane is approximately formulated as

$$L = \mu_o \frac{l}{2\pi}\left[\ln\left(\frac{2l}{w+t}\right) + 0.5 + \left(\frac{0.22(w+t)}{l}\right)\right] \tag{2.62}$$

$$M_{sg} = \mu_o \frac{l}{2\pi}\left[\ln\left(\frac{2l}{dsg}\right) - 1 + \frac{dsg}{l}\right] \tag{2.63}$$

$$M_{gg} = \mu_o \frac{l}{2\pi}\left[\ln\left(\frac{2l}{dgg}\right) - 1 + \frac{dgg}{l}\right] \tag{2.64}$$

where $\mu_o = 4\pi 10^{-7}$ H/m is permeability, M_{sg} is magnetic inductance between the center of individual signal net and ground shield and M_{gg} is between the centers of two ground shields.

High-Frequency Effects: Skin Effect

At high frequencies, the current density in an interconnect wire is no longer uniform. Due to the phenomenon called *skin effect,* the current tends to flow near the interconnect surface. This skin effect reduces the effective cross-sectional area of thick conductors and raises the effective resistance at a high frequency. The skin depth for a conductor is the distance below the conductor surface where the current density drops to $1/e$ of that at the surface, and is determined as

$$\delta(f) = \sqrt{\frac{\rho}{\pi \mu f}} \tag{2.65}$$

where μ is the magnetic permeability of the dielectric in free space. It has been observed that as the frequency increases to tens of GHz, the skin depth enters the DSM region and decreases slowly. In a chip with a good power grid, good current return paths are usually available on all sides. Thus, it is a reasonable approximation to assume the current flows in a shell of thickness δ along the four sides of the conductor, as shown in Figure 2.29.

Whether to consider these non-ideal effects depends upon the accuracy requirements of the models and the operating regime of the circuits. Often, more than one effect needs to be simultaneously considered. For example, the skin effect and surface scattering effect when simultaneously considered is known as the anomalous skin effect (ASE).

2.5.2 Geometry of CNT Interconnect and Impedance Calculation

In order to study the electrical behavior of carbon nanotubes as interconnecting wires and also to juxtapose their performance with conventional Cu interconnects, it is necessary to postulate a model describing the electromagnetic field propagation. To date, three theories have been posit to postulate different electrical equivalent models describing the basic geometry of CNT conducting wires.

In 2002, Burke (2002, 2003) stated that electrons are strongly correlated when they transport along the CNT and thus became a pioneer by first modeling a nanotube as a nano-transmission line with distributed quantum and electrostatic capacitance as well as magnetic and kinetic inductance using Lüttinger liquid collective modes. The Lüttinger liquid theory (Fisher and Glazman, 1996) describes interacting electrons (or other fermions) in a one-dimensional conductor and is necessary since the commonly used Fermi liquid model breaks down in one dimension. To posit and further study the models, subsequently the Lüttinger liquid model was also explored by various other researchers, namely Avouris et al. (2003),

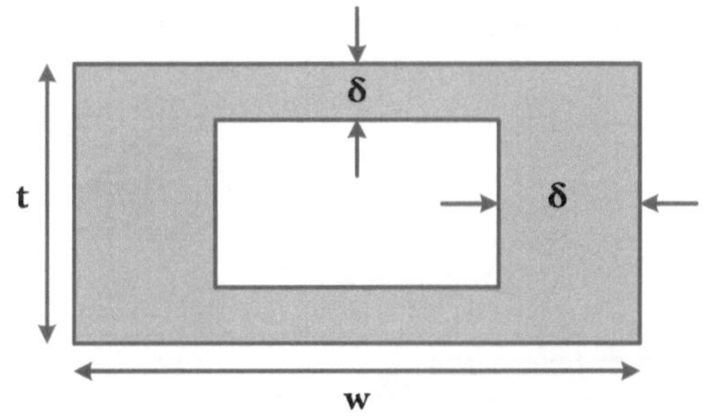

FIGURE 2.29 Current flow determined by skin depth.

Li et al. (2003), Miano and Villone (2005), Maffucci et al. (2009) and Salahuddin et al. (2005).

Another transmission line model that was put up next was based on the Boltzmann transport equation (BTE) [86] (Salahuddin et al., 2005). Two-dimensional electron gas, where the charged particles are confined to a plane and neutralized by an inert uniform rigid positive plane background was studied by Fetter (1973). Based on the extended liquid theory and the work of Fetter, a third model called the fluid model was proposed by Maffucci et al. (2008) after investigating electron transport along the CNT. This transmission line (TL) model presented by Maffucci describes the propagation along SWCNT interconnects for both isolated and bundled CNTs. Using this TL model, for a small CNT radius, all the per unit length (p.u.l.) parameters were considered similar to models presented in Miano and Villone and Salahuddin et al. Before moving ahead to the next sections that describe in detail the geometry of different types of CNT interconnects and equivalent RLC model having p.u.l parameters, it can be summarized that among the three main models postulated, the first model is based on quantum dynamics concepts, the second model requires solving the Boltzmann transport equation and the third model has been developed within the framework of classical electrodynamics and is simple in concepts and mathematical modeling.

Equivalent RLC Model of CNT Interconnect

This section presents an equivalent electrical model of SWCNT, MWCNT and SWCNT bundle interconnects. Depending on the geometry, distributed RLC parameters and their equivalent circuit models are as follows.

Single-Walled CNT (SWCNT) Interconnect

Using the Landauer–Buttiker (Imry and Landauer, 1999) formalism, a single-walled carbon nanotube (SWCNT) is modeled as a 1D quantum wire (Iijima, 1991) and in accordance to the Luttinger liquid collective modes, an isolated (SWCNT) cylinder of diameter d is placed at a distance y from the ground plane, as shown in Figure 2.30.

P. J. Burke introduced an equivalent electrical circuit of SWCNT interconnect stemmed from Luttinger's liquid theory, formulating CNTs as tandem of a series RLC nano-transmission line with quantum and electrostatic capacitance as well as distributed kinetic and magnetic inductance shown in Figure 2.31.

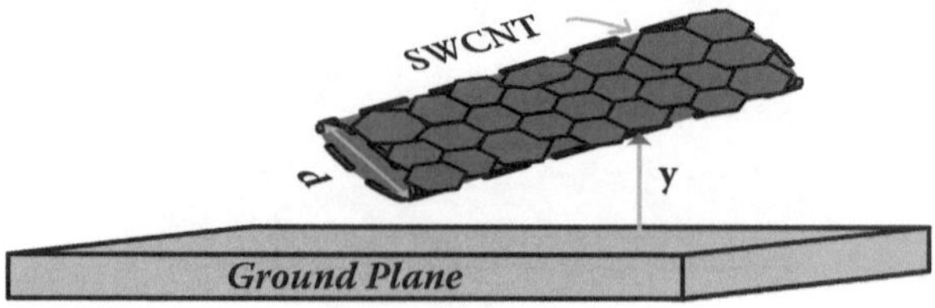

FIGURE 2.30 SWCNT conductor model.

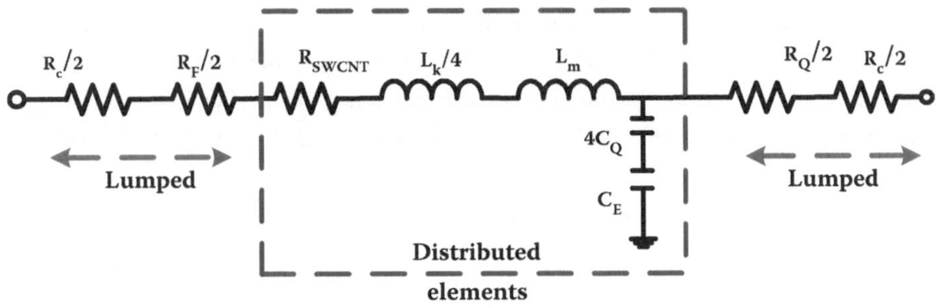

FIGURE 2.31 TLM of an isolated SWCNT.

The analytical expressions for calculating RLC parameters of a CNT interconnect are recapitulated as follows.

Resistance of Isolated SWCNT

The total effective resistance manifested by an isolated SWCNT is mathematically expressed as

$$\boldsymbol{R_{CNT} = R_C + R_F + R_{swcnt}} \tag{2.66}$$

The contact resistance, $\boldsymbol{R_C}$, arises because of scattering phenomenon at the contacts when interfaces are connected in a poor fashion. In tubes with small diameters, this undesirable effect becomes more prominent because of poor and imperfect bonding between nanotubes and metal. The consequence of poor bonding is chocking of current on the edge of a metal contact, leading in a non-homogeneous current flow from SWCNT to metal. The preferred $\boldsymbol{R_C}$ value taken into consideration for the simulation purpose is 40.7 + 0.7 kΩ.

Fundamental or quantum resistance associated with SWCNT that is equally divided between two contacts on either side of the tube is mathematically expressed as

$$\boldsymbol{R_F = \frac{h}{Ne^2T}} \tag{2.67}$$

where T is a transmission coefficient for electrons and assuming perfect contacts T = 1. N is the channel in parallel. Equation (2.67) is widely known as Landauer's formula after Rolf Landauer, who pioneered the development of the expression. As a consequence of sub-lattice degeneracy and spin degeneracy of electrons in graphene, each nanotube has conventionally $N = 4$ *channels*. N is the only variable in the quantum conductance representing the material, temperature and field dependence. The adaption of this convention can be made more vivid with an understanding of the concept that due to extremely small diameters of nanotubes, the quantization in CNTs is in circumferential direction. Hence, it creates a large energy spacing in between 1D sub-bands. Thus, if an electron jumps from a 3D metal contact to a 1D single-walled nanotube it encounters near its Fermi energy, only two are sub-bands (present because of orbital degeneracy). As each sub-band has the quantum resistance of $\frac{h}{e^2}$, thus combined resistance of an electron will be $\frac{h}{2e^2}$. Furthermore, each sub-band can lodge two electrons of different spins. Therefore, total fundamental resistance becomes $R_F = \frac{h}{4e^2}$. Substituting values of h and e, R_F comes out to be 6.45 kΩ, which is the least resistance that an electron will face on entering SWCNT. For lengths shorter than the mean free path of electrons in CNT, the resistance will be equal to fundamental resistance due to the ballistic nature of electron transport within the nanotube, but if length exceeds its mean free path resistance, it becomes directly proportionate function of length and the formula changes to

$$R_{SWCNT} = \left[\frac{h}{4e^2}\right]\left[\frac{L}{L_o}\right] \tag{2.68}$$

where L is length of CNT and L_o is the mean free path (defined as the distance across which no scattering occurs), which is 1 μm for CNT. The contact resistance and quantum resistance are independent of CNT length and are therefore modeled by two lumped values, $R_C/2$ and $R_Q/2$, at both of the ends.

Capacitance of Isolated SWCNT

CNT overall capacitance comprises a series combination of electrostatic capacitance, C_E, and quantum capacitance, C_Q. Electrostatic capacitance is per unit length of nanotube and is mathematically expressed as

$$C_E = \frac{2\pi\varepsilon}{\ln\left(\frac{y}{d}\right)} \tag{2.69}$$

It is mainly due to the storage charge by the CNT ground plane system. Quantum capacitance accounts for energy stored in a nanotube when it carries current and is given by

$$C_Q = \frac{2e^2}{hv_f} \tag{2.70}$$

where v_f is the Fermi velocity, 8×10^5 m/s and C_Q is ≈ 100 aF/μm. Since SWCNT has $N = 4$ conducting channels, the total effective capacitance is $4C_Q$.

Inductance of Isolated SWCNT

Effective SWCNT inductance consists of two components: the per unit length magnetic inductance, L_M (equation), generated due to the magnetic field around a current-carrying conductor in the presence of a ground plane and kinetic inductance, L_K, because of the kinetic energy of electrons.

$$L_M = \mu \ln\left(\frac{y}{d}\right) = \frac{\mu}{2\pi}\cosh^{-1} y/d \tag{2.71}$$

$$L_K = \left(\frac{h}{2e^2 v_f}\right) \tag{2.72}$$

If d = 1 nm and y = 1 μm, the value of L_M = 1.4 pH/μm and L_K = 16 nH/μm. It is important to note that for an isolated CNT, magnetic inductance is insignificant and the kinetic inductance dominates. Since $L_K \gg L_M$, inclusion of L_M does not have a significant impact on the delay model for isolated SWCNT interconnects. As there are four conducting channels in a CNT, the effective kinetic inductance of an isolated CNT is $L_K/4$.

From the equation we can interpret that an isolated SWCNT has a very large value of ($\boldsymbol{R_C}$) contact resistance. This factor hinders the application of isolated SWCNT for next-generation interconnects (Srivastava et al., 2010), whereas MWCNT and CNT bundles give low contact resistance when used as the circuit interconnects (Nihei et al., 2005; Massoud and Nieuwoudt, 2006).

A one-dimensional fluid model, which has been applied in modeling of an isolated SWCNT interconnect, can also be extended in modeling multi-walled carbon nanotube interconnect and SWCNT bundle interconnects, which is described in the following sections.

Multi-Walled CNT (MWCNT) Interconnect

Multi-walled carbon nanotube interconnects, because of their good current-carrying capabilities, have recently gathered the attention of many researchers as a suitable alternate material for modeling on-chip interconnects in next-generation integrated circuits.

MWCNT is a carbon nanotube structure with many concentric shells of rolled graphene sheets. The number of shells in MWCNT is diameter dependent. MWCNTs can be broadly classified as (1) double-walled CNT (DWCNT) and (2) multi-walled CNT (MWCNT). DWCNT and MWCNT have two and many concentric shells, respectively.

To analyze their electrical behavior, Li et al. (2008) proffered a multi-conductor transmission line (MTL) model for the MWCNT. In this work, they considered the tunneling effect between of the adjacent shells in MWCNT and neighboring CNTs in a bundle. However, using the MTL model, the analysis of MWCNT with *N* number of tubes leads to the solution of differential equations with the system dimensional of 2N, which can be computationally expensive. For this reason, the equivalent single conductor (ESC) model was proposed in Sarto and Tamburrano (2010). The ESC model is based on the assumption that voltages at an arbitrary cross section along MWCNT are the same, such that all nanotubes are connected in parallel at both ends. The accuracy of the ESC model in comparison to the MTL model has been reported by several researchers. It was observed that the transient responses to a pulse input of the MTL model and ESC model are in good agreement. The MTL and ESC models are briefly described in the next section.

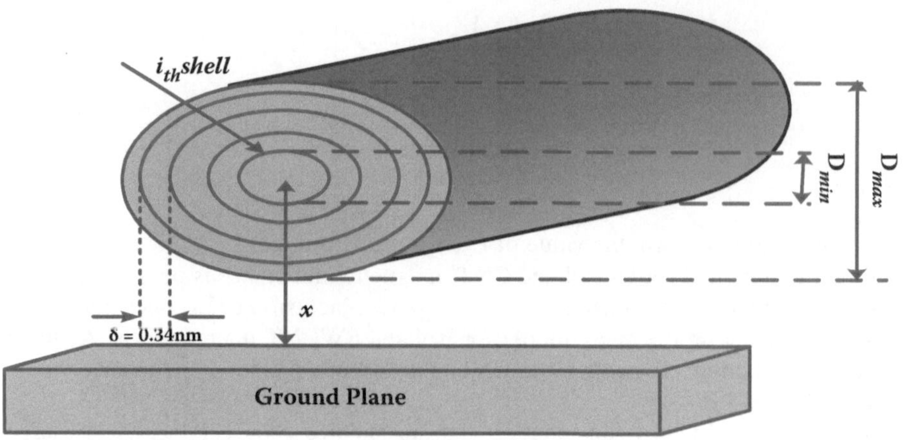

FIGURE 2.32 MWCNT conductor model.

MTL and ESC Models of MWCNT Interconnects As shown in Figure 2.32, a schematic cross-sectional view of an MWCNT interconnect is positioned over a ground plane at a distance x and placed in a dielectric medium with dielectric constant ε. The MWCNT interconnect consists of N number of tubes with an intershell distance δ = 0.34 nm, which is the van der Waal gap. The outermost shell diameter is D_{max}, and the innermost shell diameter is D_{min}. The total number of CNT shells in an MWCNT can be expressed as

$$N = 1 + int\left[\frac{(D_{max} - D_{min})}{2\delta}\right] \quad (2.73)$$

where int[·] represents an integer value. If the shells are counted from the outer to the inner as 1, 2,.. k, …N, the diameter of the k_{th} shell is given by

$$D_k = D_{max} - 2\delta(k-1) for\ 1 \le k \le N \quad (2.74)$$

The number of conducting channels in a CNT can be derived by adding all the sub-bands contributing to the current conduction. Using the Fermi function, it can be calculated as

$$N_{ch,i} = \sum_{subbands} \frac{1}{exp(|E_i - E_F|/k_B T) + 1} \quad (2.75)$$

where T is the temperature, k_B is the Boltzmann constant and E_i is either the lowest or highest energy for the sub-bands above or below the Fermi level, E_F.

The multiconductor transmission line (MTL) model of MWCNT interconnect is shown in Figure 2.33 and its equivalent single conductor model is shown in Figure 2.34.

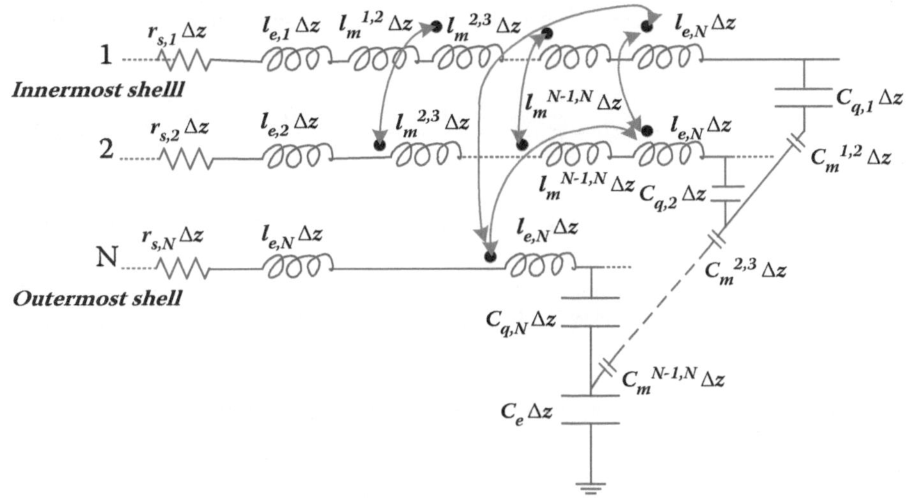

FIGURE 2.33 The multiconductor transmission line (MTL) model of MWCNT interconnect.

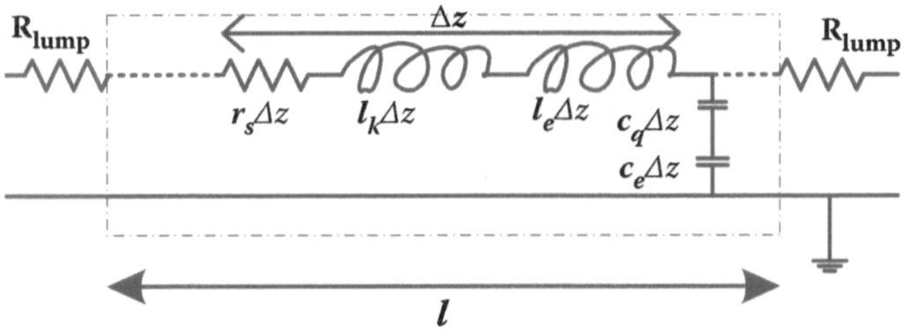

FIGURE 2.34 The electrical equivalent single conductor model of MWCNT interconnects.

The RLC parasitic of an MWCNT interconnect is primarily dependent on the number of conducting channels and can be extracted as follows.

Resistance of Individual Shell of MWCNT The resistance of any k_{th} shell consists of three parts: the first one is the quantum contact resistance R_Q, which is due to the finite conductance value of quantum wire if there is no scattering along the length; the second one is the scattering-induced ohmic resistance, $r_{s,k}$, which is due to acoustic phonon scattering and optical phonon scattering that occurs when the nanotube lengths exceed the mean free path of electrons and the third one is imperfect metal nanotube contact resistance, R_{MC}. Scattering resistance, $r_{s,k}$, occurs only if the length of the nanotube is larger than the electron MFP; it is measured per unit length and is distributed along the length of CNT, while quantum resistance, R_Q, and contact resistance, R_{MC}, appear as lumped parameters and are considered at the contacts of near-end and far-end terminals. R_Q and $r_{s,k}$ are intrinsic, and R_{MC} is due to the fabrication process. The value of the intrinsic resistance of each shell is determined by

$$R_{shell}^{k} = R_{Q}^{k} + r_{s}^{k} = \frac{h}{4e^2 N_{ch,k}} + \frac{h}{2e^2 \lambda_{mfp,k} N_{ch,k}} \tag{2.76}$$

where h, e and λ_{mfp} denote Planck's constant, electron charge and mean free path, respectively.

The imperfect contact resistance, R_{MC}, can range from 0 to 100 kΩ for different growth processes. R_{MC} in MWCNT could be very small compared to the total resistance.

Inductance of Individual Shell of MWCNT The MWCNT demonstrates two different types of inductances.

The p.u.l. kinetic inductance of the kth shell is mainly due to the kinetic energy of electrons. By equating kinetic energy stored in each conducting channel of a CNT shell to the effective inductance, the kinetic inductance of each conducting channel in a CNT can be given by

$$L_{k,k} = \frac{h}{4e^2 \, v_F N_{ch,k}} \tag{2.77}$$

The p.u.l. magnetic inductance of each shell is negligible as compared to kinetic inductance and is due to the magnetic field generation around a current-carrying conductor. In the presence of a ground plane, the p.u.l. magnetic inductance of a CNT shell is expressed as

$$L_{e,k} = \frac{\mu_0 \mu_r}{2\pi} \cosh^{-1}\left[\left(\frac{d_k + 2x}{d_k}\right)\right] \tag{2.78}$$

The mutual inductance between the shells is mainly due to the magnetic field coupling between the adjacent shells in an MWCNT. The p.u.l. mutual inductance can be expressed as

$$L_{m}^{k,\,k+1} = \frac{\mu}{2\pi} \ln\left(\frac{d_{k+1}}{d_k}\right), \; k = 1, 2, \ldots \ldots, \mathrm{N} - 1 \tag{2.79}$$

Capacitance of Individual Shell of MWCNT The MWCNT interconnect consists of two types of capacitances. The first one is the electrostatic capacitance, C_e, due to electrostatic field coupling between the CNT and the ground plane. The electrostatic capacitance of MWCNT appears between the outermost shell and the ground plane, as the outer shell acts as a shield for the internal ones. The p.u.l. C_e of the CNT outermost shell shown can be expressed as

$$C_e = \frac{2\pi\varepsilon}{\cosh^{-1}\left(\frac{d+2x}{d}\right)} \tag{2.80}$$

The p.u.l. electrostatic mutual capacitance, C_m, is mainly due to the potential difference between adjacent shells in MWCNT. The intershell coupling capacitance between the k_{th} and $(k + 1)_{th}$ shells is given by

$$C_m = \frac{2\pi\varepsilon}{\ln\left(\frac{d_{k+1}}{d_k}\right)} \tag{2.81}$$

Second is the quantum capacitance that originates from the quantum electrostatic energy stored in a CNT shell when it carries current. According to the Pauli exclusion principle, it is only possible to add extra electrons into the CNT shell at an available state above the Fermi level. By equating this energy to the effective capacitance energy, the quantum capacitance of each conducting channel in a CNT can be expressed as

$$C_q = \frac{2e^2}{\mathrm{h}v_F} \tag{2.82}$$

To reduce the complexity of the MTL model, a simplified ESC model was proposed. The ESC model is shown in Figure 2.34. This model was developed based on the assumption that voltages at an arbitrary cross section along the MWCNT are the same. Thus, all the scattering resistances, $r_{s,k}$, are in parallel and can be replaced by an equivalent resistance, $r_{s,ESC}$. The $r_{s,ESC}$ can be expressed as

$$r_{s,ESC} = \frac{h/e^2}{\sum_{k=1}^{N} 2N_{ch,k}\,\lambda_{mfp,k}} \tag{2.83}$$

The distributed equivalent MWCNT capacitance, $c_{q,ESC}$, in terms of quantum capacitance and coupling capacitance between shell to shell is expressed as

$$c_{q.ESC} = c_{equ.N} \tag{2.84}$$

$$c_{equ.k} = \left(\left(\frac{1}{c_{equ,k-1}} + \frac{1}{c_m^{k-1,k}}\right)^{-1}\right)$$

$c_m^{k-1,k} = \frac{2\pi\varepsilon}{ln\,(d_{k+1}/\,d_k)}$ for k = 2,3, …., N

By adopting a recursive approach proposed in Sarto and Tamburrano (2010), the equivalent inductance can be expressed as

$$l_{equ,k} = l_{k,ESC} = \left(\left(\frac{1}{l_{equ,k-1} + l_m^{k-1,k}} + \frac{1}{l_{k,i}} \right)^{-1} \right) \tag{2.85}$$

SWCNT Bundle Interconnect

Single-walled carbon nanotubes can also be fabricated as a bundle, which means CNTs in a bundle are parallel to each other. The spacing between CNTs in the bundle is due to the van der Waals forces between the atoms of neighboring nanotubes (Xu et al., 2009). One of the most critical challenges in realizing high-performance SWCNT-based interconnects is controlling the proportion of metallic nanotubes in the bundle. Current SWCNT fabrication techniques cannot effectively control the chirality of the nanotubes in the bundle. Therefore, SWCNT bundles have metallic nanotubes that are randomly distributed within the bundle. The conventional one-dimensional fluid model described in the preceding section can be applied to each individual SWCNT in the bundle with some modification, as the electron–electron interaction in the SWCNT bundle is different from that in an isolated SWCNT.

For analytical purposes, we consider a single SWCNT in a bundle and assume that the electrons in this SWCNT are solely affected by the electrons in the neighboring metallic SWCNTs and semiconducting SWCNTs have no effect on the conductance of the bundle.

As shown in Figure 2.35, the schematic cross-sectional view of the SWCNT bundle interconnect has height h and width w. The bundle contains a number of SWNTs of diameter D. The center-to-center distance $S_{c\text{-}c}(=D+\delta)$ between adjacent SWCNTs in a bundle is $S_{C-C} = (D + \delta)$, where $\delta = 0.34$ nm represents the van der

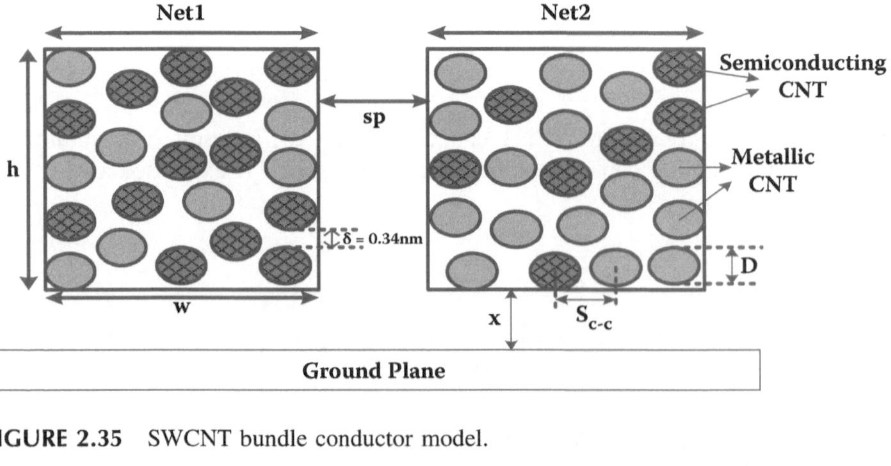

FIGURE 2.35 SWCNT bundle conductor model.

Waal's distance between adjacent carbon atoms. The total number of SWCNTs (N_{SWCNT}) in a bundle can be calculated as

$$N_{SWCNT} = \begin{cases} n_{rows}n_{column} - (n_{rows}/2) \ \textit{if number of rows areeven} \\ n_{rows}n_{column} - (n_{rows} - 1)/2 \ \textit{if number of rows are odd} \end{cases} \tag{2.86}$$

$$n_{rows} = \left[\frac{h - D}{S_{c-c}}\right] + 1; \ n_{column} = \left[\frac{w - D}{\sqrt{3}/2S_{C-C} + 1}\right] \tag{2.87}$$

It was mentioned previously that due to the lack of control on chirality, a bundle consists of both metallic and semiconducting nanotubes. Only the metallic ones contribute to current conduction, and hence, the effective number of conducting channels in a SWCNT bundle with both metallic and semiconducting CNTs is less than that of a SWCNT bundle with all metallic CNTs. A bundle with all metallic CNTs is termed a densely packed bundle, and a bundle with both metallic and semiconducting CNTs is termed a sparsely packed bundle. For modeling a sparsely packed bundle, it is advised to assume larger inter CNT-CNT spacing ($S_{c-c} > D + \delta$).

Depending on the bundle geometry, an equivalent RLC model of the SWNT bundle is considered. This section presents a brief description of equivalent SWCNT bundle resistance, inductance and capacitance.

Resistance of SWCNT Bundle interconnect The resistance of a SWCNT bundle consisting of a (N_{SWCNT}) number of SWCNTs is given as

$$R_{bundle} = \frac{R_{SWCNT}}{P_m N_{SWCNT}} = \frac{(R_c + R_q)}{P_m N_{SWCNT}} + \frac{R_q l_{SWCNT}}{P_m N_{SWCNT} \lambda_{mfp}} = R_{i-b} + r_{d-b} l_{SWCNT} \tag{2.88}$$

where P_m is the fraction of metallic tubes in a bundle and R_{i-b} and r_{d-b} denote the lumped and distributed resistances of the SWCNT bundle, respectively.

Inductance of SWCNT Bundle interconnect The kinetic inductance of the bundle is expressed as

$$L_{kinetic\_bundle} = l \ \frac{L_K/4}{P_m N_{SWCNT}} \tag{2.89}$$

the magnetic inductance of the SWCNT bundle is premeditated by taking into consideration the mutual inductances between the SWCNTs in a bundle using the partial inductance modeling approach based on the partial element equivalent circuit (PEEC) method. The partial self-inductance ($L_{magnetic}$) of an isolated SWCNT is given by Nieuwoudt and Massoud (2006)

$$L_{magnetic} = \frac{\mu_0 l}{2\pi}\left[\ln\frac{l}{D} + \frac{1}{2} + \frac{2D}{3l}\right] \tag{2.90}$$

The mutual inductance (M_m)between two parallel SWCNTs of equal length is given by

$$M_m = = \frac{\mu_0 l}{2\pi}\left[ln\rho + \sqrt{1+\rho^2} - \sqrt{1+\frac{1}{\rho^2}} + \frac{1}{\rho}\right] \tag{2.91}$$

where $\rho = \frac{1}{S_{c-c}}$. The overall loop inductance of the bundle calculated in accordance to the PEEC approach is expressed as

$$L^M_{bundle} = i_n^T L_{mat} i_n \tag{2.92}$$

where i_n is a vector with normalized current in each SWCNT.

L_{mat} is the partial inductance matrix given by

$$L_{mat} = \begin{bmatrix} L_m^1 & M_m^{1,2} & \cdots & M_m^{1,n} \\ M_m^{2,1} & L_m^2 & \ldots & M_m^{2,n} \\ \vdots & \vdots & \ddots & \vdots \\ M_m^{n,1} & M_m^{n,2} & \cdots & L_m^n \end{bmatrix}$$

where L_m^j is the partial self-inductance of the j_{th} SWCNT and $M_m^{i,j}$ is the partial mutual inductance between the i_{th} and j_{th} SWCNTs.

The total equivalent inductance of a SWCNT bundle is the series combination of magnetic and kinetic inductances and can be given as

$$L_{bundle} = L^M_{bundle} + L^K_{bundle} \tag{2.93}$$

Capacitance of SWCNT Bundle Interconnect The quantum capacitance of a SWCNT bundle mainly depends on the total number of SWCNTs and can be expressed as

$$c_{q-b} = c_q.\ N_{SWCNT} \tag{2.94}$$

As already mentioned, due to the four conducting channels of SWCNT, the equivalent quantum capacitance of the SWCNT bundle interconnects is equal to $4c_{q-b}$. The electrostatic capacitance of the SWNT bundle can be obtained as (Pu et al., 2009)

$$c_{e-b} = \frac{2\pi\varepsilon_0\varepsilon_r}{ln\,(x/D)} \times n_x \tag{2.95}$$

where n_x denotes number of SWCNTs facing the ground plane.

2.5.3 Geometry of GNR Interconnect and Impedance Calculation

The analytical expressions for calculating the RLC parameters of the MLGNR interconnect are recapitulated below.

The total count of the layers in an MLGNR can be determined with the help of the following mathematical expression:

$$N = 1 + \boldsymbol{Integer}\left(\frac{\tau}{\delta}\right) \tag{2.96}$$

Every layer of MLGNR is made up of several conducting channels denoted by the variable N_{ch}. The effective number of conducting channels plays a vital role in controlling the MLGNR interconnects' performance as its parasitic impedance (RLC) parameters are a function of N_{ch}. In order to commensurate the value of N_{ch}, we have to take into consideration the Fermi energy (E_F), temperature (T), width ω and also the impact of spin and sublattice degeneracy of carbon atoms

$$\boldsymbol{N_{ch}} = \sum\nolimits_{n=0}^{n_c} \left[\boldsymbol{e}^{(E_i-E_F)/kT} + 1\right]^{-1} + \sum\nolimits_{n=0}^{n_v} \left[\boldsymbol{e}^{(E_i+E_F)/kT} + 1\right]^{-1} \tag{2.97}$$

where k, n_c and n_v denote the Boltzmann constant, number of conduction bands and number of valence bands, respectively. The term $\boldsymbol{E_i}$ delineates the lowest/highest energy of the ith conduction/valence sub-band and can be calculated as

$$\boldsymbol{E_i} = \begin{cases} \frac{3\sqrt{3}\,\boldsymbol{a_0\,ct}}{2(\omega + \sqrt{3}\,\boldsymbol{a_0})}, & n = 0 \\ \frac{\boldsymbol{h\,.\,v_f}}{2\,.\,\omega}|\boldsymbol{n}|, & n \neq 0 \end{cases} \tag{2.98}$$

where the distance between two carbon-carbon atoms is denoted by a_0 ($\approx 0.142\ nm$), c (= 0.12), t (= $2.7eV$), h is Planck's constant and Fermi velocity is v_f (= 8×10^5) m/s.

Resistance of MLGNR interconnect

The equivalent resistance of the MLGNR interconnect has two resistive components called lumped resistance (R_{lumpd}) and per unit length distributed resistance ($R_{distributd}$)

$$(\boldsymbol{R_{lumpd}}) = \frac{1}{\{\sum_{i=1}^{N}(\boldsymbol{R_c^i} + \boldsymbol{R_q^i})^{-1}\}} \tag{2.99}$$

where contact resistance is due to imperfect contact between the metal layer and GNR layers is denoted by R_c^i (has values ranging from 1 to 20 kΩ depending on the fabrication process) and quantum resistance due to quantum effects of charge carriers in MLGNR layers is denoted by $\boldsymbol{R_q^i}\left(\approx \frac{h}{2e^2 N_{ch}} \boldsymbol{k\Omega}\right)$. The integer i has values for a number of layers ranging from 1 to N.

While the distributed resistance ($R_{distributd}$) , because of phonon and acoustic scattering in MLGNR, is expressed as

$$(\boldsymbol{R_{dis}}) = \frac{\boldsymbol{R_s^i}}{\{N\}} \tag{2.100}$$

$$\boldsymbol{R_s^i} = \frac{\boldsymbol{h}}{2e^2.\ N_{ch}.\ \lambda_{eff}} = \frac{12.9}{N_{ch}.\ \lambda_{eff}}\boldsymbol{K\Omega} \tag{2.101}$$

where e denotes electronic charge and λ_{eff} denotes the effective mean free path (MFP) of electrons. The MFP is considered a vital parameter in order to determine the rate of conduction of current in the case of multi-layered graphene nanoribbon (MLGNR) interconnects. In general, the MFP for MLGNR with smooth edges is approximately taken to be around ≈ 1 μm, but in the case of MLGNR with rough edges, the MFP is shorter. Here in this typescript for impedance analysis, two mean free paths are taken into consideration. The combinations of these two are termed effective MFP (λ_{eff}). The two different MFPs are determined considering edge scattering (including diffusive scattering and surface-induced disorder) and another is due to electron-electron scattering, acoustic phonon scattering and remote interfacial phonon scattering.

Applying Matthiessen's rule to calculate MFP for the nth sub-band we get

$$\lambda_{eff,n} = \frac{1}{\frac{1}{\lambda_d} + \frac{1}{\lambda_n}} \tag{2.102}$$

where λ_d MFP is due to various types of scattering phenomenon of electrons (viz acoustic, optical phonon, impurity and defect scattering). λ_n is an MFP due to edge scattering and can be computed as

$$\lambda_n = \frac{\omega}{1 - \boldsymbol{P}}\sqrt{\left(\frac{2.\ \omega.\ \boldsymbol{E_f}}{\boldsymbol{n}.\ \boldsymbol{h}.\ \boldsymbol{v_f}}\right)^2 - 1} \tag{2.103}$$

where P (varies 0 to 1) is a specular constant and determines the extent of edge roughness in GNRs.

Inductance of MLGNR Interconnect

The equivalent inductance consists of two components: kinetic inductance L_K and magnetic inductance L_m

$$L = L_K + L_m \tag{2.104}$$

The kinetic inductance, L_K, is due to the inertia of a mobile charge carrier of each sheet of MLGNR and is expressed as

$$L_k = \frac{L_{k_0}}{2N_{ch}}; \text{ Where } L_{k_o} = \frac{h}{2e^2 v_F} \tag{2.105}$$

where v_F is the Fermi velocity ($\approx 8 \times 10^5 \frac{m}{s}$) of carriers in graphene. The value of (L_k) per channel is experimentally calculated as 8 nH/μm.

In addition to this, because of stored energies of carriers in the magnetic field, the MLGNR wires exhibit a magnetic inductance denoted by L_m and expressed as

$$L_m = \frac{\mu_o \mu_r d}{w} \tag{2.106}$$

Capacitance of MLGNR Interconnect

Each layer of interconnect has a quantum capacitance, c_q, which is a function of density of electronic states and given by

$$c_q = 2c_{q_0} \,.\, N_{ch} \text{ where } c_{qo} = \frac{2e^2}{hv_F} \tag{2.107}$$

As a result of electric field coupling between the ground plane and bottom-most layer of MLGNR, it introduces an electrostatic capacitance, c_e, which is a function of the width, w, of nanoribbons and the distance, d, from the ground and is given by

$$c_e = \frac{\varepsilon_o \varepsilon_r w}{d} \tag{2.108}$$

The mutual inductance, l_m, and coupling capacitance, c_m, occurring in between adjacent GNR sheets are expressed as

$$l_m = \frac{\mu_o \delta}{w} \text{ and } c_m = \frac{\epsilon_o w}{\delta} \tag{2.109}$$

TABLE 2.6
ITRS defined interconnect technology parameters

Year	2010	2013	2016	2019
DRAM half pitch (nm)	45	32	22	16
MPU M1 half pitch (nm)	45	32	22	16
MPU gate length (nm)	18	13	9	6
No. of metal level (nm)	12	13	13	14
M1 wiring pitch (nm)	90	64	44	32
M1 A/R	1.8	1.9	2	2
Effective resistivity (μΩ-cm)	4.08	4.83	6.01	7.34
V_{DD}(V)	1.0	0.9	0.8	0.7
Dielectric constant	2.5-2.8	2.1-2.4	1.9-2.2	1.6-1.9

Interconnect Technology and Impedance Parameters

The interconnect parameters taken from ITRS (Wilson, 2006) are tabulated below. The CNIA tool is used for extracting RLC parameters of interconnects and also equations defined in Section 2.5 are used for computing values of RLC under different scenarios (Table 2.6).

From the impedance values tabulated for copper (Cu) interconnect wires at different technology nodes starting from 45 nm to 16 nm in Table 2.7, we infer that with the advancement in technology nodes the impact of resistance becomes prominent, whereas the inductance values do not show much variation. We may state that scaled interconnect wires suffer from a hike in resistance due to a reduction in cross-section area of a chip; moreover, if the conductor height is not proportionately reduced with spacing so as to maintain a proper aspect ratio, then capacitance also increases. The impedance (RLC) parameters play a vital role in determining the latency and power consumption effect of on-chip interconnects. Analyzing the values of Table 2.8 for isolated singled-walled carbon nanotubes and comparing them with Table 2.7, we conclude that the Cu interconnect wire offers a large value of resistance as compared to SWCNT interconnects. It is also evident that for a shorter wire length, the impedance is less at all technology nodes for both of the materials. With an increase in length as we move from local interconnects to intermediate and global interconnects, the resistance increases abruptly. The single-walled nanotubes of carbon exhibit a conductivity higher than that of copper; because of the fact that juxtapose to mean free path of copper, which is only of the order of a few tens of a nanometer, the mean free path of SWCNT is several micrometers long. In addition to that, it is proven that SWCNT wires are able to carry a current density of the order of 10^{14} A/m^2 or greater. A single-layer nanotube interconnect wire offers 50% lesser capacitance in comparison to copper interconnect wires and thus can save a significant amount of power. In general, local interconnects are the chief source of energy dissipation. Moreover, dynamic delay variation because of different switching patterns is considerably mitigated due to

TABLE 2.7
Length-dependent RLC parameters at distinct technology nodes for Cu interconnect

Resistance (kΩ)	Length (μm)	45 nm	32 nm	22 nm	16 nm
	1	0.011	0.026	0.059	0.144
	5	0.055	0.123	0.33	0.707
	10	0.10	0.26	0.63	1.44
	50	0.57	1.23	3.01	7.18
	100	1.13	2.49	6.21	14.35
Inductance (pH)	1	0.66	0.72	0.79	0.85
	5	4.87	5.20	5.52	5.88
	10	11.15	11.75	12.43	13.07
	50	71.75	74.83	78.25	81.43
	100	157.4	163.53	170.32	176.6
Capacitance to ground (fF)	1	0.028	0.021	0.018	0.017
	5	0.132	0.109	0.095	0.081
	10	0.267	0.216	0.189	0.164
	50	1.329	1.081	0.940	0.813
	100	2.66	2.151	1.884	1.626
Coupling capacitance (fF)	1	0.069	0.062	0.058	0.050
	5	0.341	0.307	0.293	0.252
	10	0.673	0.611	0.576	0.507
	50	3.412	3.056	2.923	2.524
	100	6.824	6.113	5.845	5.048

maximum variation in wire capacitance for SWCNT of 80% contrary to 200% for Cu wires.

Table 2.9 shows the length-dependent RLC parameters' value at distinct technology nodes for SWCNT bundle interconnects. As the resistance of an isolated singled-walled nanotube of carbon is quite high, of the order of 6.45 kΩ, hence in order to increase the conductivity of SWCNT interconnects, it is advised to use a bundle of SWCNTs. It is interpreted from Table 2.9 that the resistance of SWCNT bundle at the local interconnect level is not varied, due to the fact that the maximum length taken for the analysis of local level interconnects is smaller than the mean free path of SWCNTs in the bundle. As a matter of fact, for the length of nanotubes less than its mean free path, the electron charge transport is primarily of ballistic nature within the nanotubes of carbon. The resistance, mainly intrinsic resistance (R_i), does not show dependency with length. At the same time, in copper wires the electrons can be backscattered by a series of small-angle scattering and the mean free path is in the range of the order of a few tens of nanometers. Hence, the resistance of Cu interconnects increase linearly with length, as tabulated. The capacitance exhibit by a SWCNT bundle is quite higher than the copper interconnect wire, thereby causing

TABLE 2.8
Length-dependent RLC parameters' value at distinct technology nodes for an isolated SWCNT interconnects.

16nm

Length (μm)	Resistance (kΩ)	Inductance (nH)	Capacitance (fF)		
			Quantum	Electrostatic	Coupling
1	0.34	0.15	10.82	0.0167	0.0464
5	0.55	0.73	54.09	0.084	0.232
10	0.79	1.45	108.19	0.169	0.465
50	2.71	7.23	540.87	0.85	2.33
100	5.11	14.45	1081.77	1.67	4.64

22 nm

Length (μm)	Resistance (kΩ)	Inductance (nH)	Capacitance (fF)		
			Quantum	Electrostatic	Coupling
1	0.218	0.1124	13.9078	0.0195	0.0538
5	0.32	0.5618	69.6	0.098	0.2685
10	0.425	1.124	138.2	0.195	0.538
50	1.338	5.617	696.5	0.99	2.686
100	2.478	11.235	1390.9	1.95	5.38

32 nm

Length (μm)	Resistance (kΩ)	Inductance (nH)	Capacitance (fF)		
			Quantum	Electrostatic	Coupling
1	0.116	0.079	20.09	0.0212	0.0597
5	0.154	0.389	100.46	0.108	0.293
10	0.197	0.777	200.98	0.213	0.587
50	0.552	3.889	1003.54	1.06	2.94
100	0.989	7.777	2009.05	2.13	5.87

45 nm

Length (μm)	Resistance (kΩ)	Inductance (nH)	Capacitance (fF)		
			Quantum	Electrostatic	Coupling
1	0.071	0.043	37.09	0.0247	0.0683
5	0.087	0.211	185.453	0.1234	0.3416
10	0.107	0.421	370.905	0.246	0.684
50	0.252	2.107	1854.53	1.236	3.416
100	0.456	4.214	3709.05	2.48	6.84

TABLE 2.9
Length-dependent RLC parameters' value at distinct technology nodes for SWCNT bundle interconnects

Parameter	*Perfect contact* (R_C = 0), *densely packed* (P_m = 1)				
Resistance (kΩ)	Length (μm)	Technology Node			
		45nm	32nm	22nm	16nm
	1	0.0112	0.0199	0.0388	0.0689
	5	0.0161	0.0278	0.0565	0.1041
	10	0.0216	0.0393	0.0784	0.1478
	50	0.0654	0.1223	0.2513	0.4875
	100	0.1203	0.324	0.4761	0.9102
		R_C= 100 kΩ, *densely packed* (P_m= 1)			
	1	0.0358	0.0666	0.1352	0.2594
	5	0.0407	0.0767	0.1541	0.2966
	10	0.0454	0.0863	0.1768	0.3382
	50	0.0903	0.1693	0.3478	0.677
	100	0.1448	0.2732	0.5546	1.1016
		R_C= 100 k Ω, *densely packed* (P_m= 1/3)			
	1	0.1077	0.1997	0.407	0.7785
	5	0.1224	0.226	0.4623	0.8867
	10	0.1392	0.2687	0.5278	1.016
	50	0.2707	0.5077	1.0464	2.0334
	100	0.4342	0.8186	1.6941	3.3053
Parameter	**Length**	**Technology node**			
		45 nm	32 nm	22 nm	16 nm
		Densely packed (P_m= 1)			
Inductance (nH)	10 (μm)	0.009	0.0187	0.0394	0.0774
Capacitance (fF)		30.34	21.07	15.20	10.45
		Sparsely packed (P_m= 1/3)			
Inductance (nH)	10 (μm)	0.0302	0.0582	0.1200	0.2356
Capacitance (fF)		24.28	6.84	12.17	8.36

greater power dissipation; but in order to optimize this it is suggested to choose an appropriate diameter of carbon nanotubes within the bundle. While considering the impedance parameters' effect on performance analysis, generally the inductance can be overlooked due to the fact that the resistive impedances are of much larger values compared to the inductive impedances at any length of interconnect. Moreover, the result analysis points out that densely packed bundles of SWCNT offer noteworthy improvements in performance in comparison to conventional Cu interconnects, despite the imperfect metal nanotube contacts. For the higher geometry and greater length, i.e. intermediate and global interconnect level, the SWCNT bundle can accommodate a larger number of nanotubes of carbon, thereby mitigating the resistance of SWCNT bundle interconnects. Hence, it can be concluded that state-of-the-art SWCNT bundles provide suitable alternatives to copper interconnects.

Table 2.10 shows the extracted R, L and C parameters for the local, intermediate and clobal interconnects of MWCNT interconnects. The RLC parasitic of an MWCNT interconnect is primarily dependent on the number of conducting channels, as illustrated in Eqs. 2.76–2.85. The resistance of any k_{th} shell consists of three parts: first is the quantum contact resistance, R_Q, which is due to the finite conductance value of quantum wire if there is no scattering along the length; the second one is the scattering-induced ohmic resistance, $r_{s,k}$, which is due to acoustic phonon scattering and optical phonon scattering that occurs when the nanotube lengths exceed the mean free path of electrons and the third is imperfect metal nanotube contact resistance, R_{MC}. Scattering resistance, $r_{s,k}$, occurs only if the length of the nanotube is larger than the electron MFP; it is measured per unit length and is distributed along the length of the CNT while quantum resistance, R_Q, and contact resistance, R_{MC}, appear as a lumped parameter and are considered at the contacts of near-end and far-end terminals. R_Q and $r_{s,k}$ are intrinsic, and R_{MC} is due to the fabrication process. Several research studies show that all shells of a MWCNT (multi-walled carbon nanotube) can conduct if they are properly connected to contacts. From the tabulated results, it can be justified that MWCNTs can have conductivities several times larger than that of Cu or SWCNT bundles for long length at a global-level interconnect. It is observed from the results that if we normalize the resistance of MWCNT with respect to SWCNT, then a decreasing trend is achieved with an increase in interconnects length, which specify that for longer interconnects (global interconnect level), the performance of MWCNT interconnects in terms of resistance is better than smaller interconnects (local interconnect level).

Recently, graphene is excessively studied because of a large mean free path of electrons and rich current-carrying ability of the order of $5\text{–}20 \times 10^8$ A/m^2. Due to the complicated fabrication procedure and non-planar nature of CNTs, it is favored to consider GNRs for the design of on-chip interconnects. On the one side, fabrication of GNRs is more controllable due to its planar nature and on the other side, further patterning of 2D GNR interconnects can be easily done using high-resolution lithography, as already discussed before. It is preferred to use multi-layer graphene nanoribbons (MLGNRs) for the interconnect applications instead of single-layer graphene. SLGNR on the one hand shows smaller values of capacitance because of their extremely smaller value of thickness, but on the other hand displays high resistance values. The benefit of using MLGNR is that it provides multiple paths for conduction of charge, henceforth mitigating the effective resistance value. From the data tabulated in Table 2.11, we interpret that as the quantum capacitance is effective in series with the electrostatic capacitance, and furthermore it increases with rise in effective number of conduction channels; hence, it can be stated that the effect of quantum capacitance is maximum at the local level interconnect for small wire widths and in single-layer graphene. Due to a lower capacitance of graphene compared to copper, it is preferred to use multi-layer graphene as interconnects in digital applications. In addition to that, with regards to their coupling with adjoining devices, multi-layered graphene nanoribbons can further be categorized as top-

TABLE 2.10
Length-dependent RLC parameters' value at distinct technology nodes for MWCNT interconnects

16nm

Length (μm)	Resistance (kΩ)	Inductance (nH)	Capacitance (fF)		
			Quantum	Electrostatic	Coupling
1	0.45	0.33	5.0225	0.0214	0.0345
5	0.60	1.55	25.112	0.1065	0.1732
10	0.77	3.12	50.225	0.2127	0.3463
50	2.19	15.54	251.13	1.0643	1.7322
100	3.86	30.12	502.25	2.1276	3.4624

22 nm

Length (μm)	Resistance (kΩ)	Inductance (nH)	Capacitance (fF)		
			Quantum	Electrostatic	Coupling
1	0.271	0.168	9.274	0.0245	0.0405
5	0.354	0.843	46.353	0.123	0.2004
10	0.432	1.684	92.735	0.245	0.401
50	1.048	8.423	463.631	1.24	2.005
100	1.836	16.851	927.261	2.47	4.01

32 nm

Length (μm)	Resistance (kΩ)	Inductance (nH)	Capacitance (fF)		
			Quantum	Electrostatic	Coupling
1	0.144	0.084	18.18	0.0267	0.0446
5	0.168	0.431	90.67	0.1344	0.2177
10	0.197	0.851	181.60	0.2688	0.4372
50	0.430	4.311	907.92	1.3443	2.1877
100	0.701	8.500	1815.78	2.6886	4.374

45 nm

Length (μm)	Resistance (kΩ)	Inductance (nH)	Capacitance (fF)		
			Quantum	Electrostatic	Coupling
1	0.078	0.046	32.87	0.0315	0.054
5	0.087	0.243	164.2	0.157	0.266
10	0.098	0.467	327.5	0.341	0.50
50	0.187	2.370	1643	1.58	2.54
100	0.289	4.768	3285	3.15	5.2

TABLE 2.11
Length-dependent RLC parameters' value at 16 nm technology nodes for GNR-based interconnects

Length (μm)	Resistance (kΩ)	Inductance		Capacitance (fF)	
		L_k (nH)	L_m (pH)	C_q (pF)	C_e (fF)
1	0.0445	0.0143	0.86	0.13	0.018
5	0.1373	0.0711	5.76	0.55	0.086
10	0.2489	0.1428	13.07	1.11	0.168
50	1.1568	0.7085	81.43	5.56	0.835
100	2.2832	1.421	176.8	11.02	1.684

contact GNRs, side-contact GNRs and side-contact MLGNRs show better performance as all the stacked graphene layers are joined physically to contact, thus offering low resistance and high conduction channels. While in the case of top-contact MLGNRs only the uppermost layer is connected with the contacts, hence equivalent resistance is high and degrades the performance of interconnects. Top-contact GNRs are analyzed by considering a 2D resistor network to find its effective resistance for different interconnect widths (Kumar et al., 2012). Unfortunately, side-contact multi-layered graphene nanoribbons with an increase in layer count turn into graphite because of interlayer electron hopping, thereby reducing its mean free path. The phenomenon of interlayer electron hopping takes place as a result of modulation in the carbon-carbon bond length, which is related to elastic strain caused by stacking of multiple layers on one another (Neto et al., 2009). In order to enhance the mean free path and the scattering rate phenomenon of GNR layers turning into graphite is averted by a technique of dielectric insertion in between GNR layers. Interlayer insertion of high-*k* dielectrics diminishes the rate of electron scattering in individual GNR layers, thereby mitigating overall resistance of multi-layered graphene nanoribbon interconnects. In the case of top-contact MLGNRs, interlayer dielectric insertion in between individual GNR layers results in marginal mitigation in the rate of electron scattering and interconnect wire resistance as in top-contact MLGNRS only the uppermost graphene layer is conducting, whereas in the case of side-contact MLGNRs all the graphene layers are conducting; henceforth the consequence of interlayer dielectric insertion in between individual GNR layers results in noteworthy mitigation in scattering rate and line resistance. Dielectric layers can be deposited using atomic layer deposition methodology. The benefit of this method is the precise control over uniformity and thickness of films. The dielectric quality sets a value of impurity concentration. The cleaner the dielectric the less the impurity concentration. The impurity concentration governs the scattering rate, which in turn determines the mean free path, charge mobility and resistance of GNR layers. Hence, the performance of MLGNR wires is indirectly impacted by impurity concentration. Thus, by choosing suitable dielectrics that are compatible with graphene, interconnect performance can be optimized.

Thermal Stability Analysis of Copper and CNT Interconnects

As discussed in former sections, conventional copper interconnects suffer from the problem of increase in resistivity primarily due to surface roughness and grain boundary scattering and electro-migration exertion on such dimensional scaling. The increased resistivity in conventional copper-based nano interconnects increases the propagation delay and power consumption. As the interconnect resistivity is a function of temperature, the propagation delay and power consumption also increase with the chip operating temperature.

Thus, the electrical behavior of high-performance nano interconnects is directly impacted by temperature fluctuations, primarily because of the effect of electron-phonon scattering on the mean free path. The mean free path plays an important role in determining the resistance of the interconnect wire. In order to study the effect of temperature on the electrical impedance parameters, Eqs. 2.55–2.57 are used to compute the values for copper interconnects. For CNT interconnects, Eqs. 2.66–2.68) are modified to compute the value of resistance as

$$R_{CNT} = \left[\frac{h}{4e^2}\right]\left[\frac{L + L_o(T)}{L_o(T)}\right] if\, L > L_o \tag{2.110}$$

$L_o(T)$ is a temperature-dependent MFP and is calculated as

$$\frac{1}{L_o(T)} = \left[\frac{T}{300L_{ac,300}} + \frac{1}{\left(L_{OP,300}\left(\frac{N_{OP}+1}{N_{OP(T)}}\right)\right)} + \frac{1}{L_{OP\mathbf{ems}}}\right]$$

where acoustic scattering length L_{ac}, 300 ≈ 1,600 nm and spontaneous optical emission length L_{OP}, 300 ≈ 15 nm.

The results obtained for resistance variations after considering the temperature-dependent electrical impedance parameters are tabulated below in Tables 2.12 and 2.13. These equations show that the inductances and capacitances of SWCNT bundle interconnects are independent from their effective MFP and therefore the impact of temperature variations on inductance and capacitance is negligible compared to the resistance and hence it is ignored in this pedagogical work.

For undergoing the thermal stability analysis of the conventional interconnect material (Cu) and then comparing the performance of interconnects with densely packed SWCNT bundles and copper wire, we calculated the resistance through interconnects of different lengths for different technology nodes. It is observed from the results tabulated in Tables 2.12 and 2.13 that with a rise in operating temperature, the effective resistance increases for both interconnect materials. Normally in the case of a SWCNT bundle, there are 33% nanotubes of carbon, which shows metallic traits while the remaining ones are semiconducting in nature. For the purpose of interconnect modeling, it is preferred to have metallic ones, as metallic nanotubes are solely responsible for the purpose of current conduction. The location

TABLE 2.12
Length-dependent resistance of Cu at different temperatures

Temperature (K)→	100	200	300	350	400	450
Length (μm) ↓			**32-nm Technology**			
10	43.14	54.76	70.47	83.27	103.11	127.53
15	47.85	66.16	89.12	107.89	136.84	175.09
20	53.54	77.56	107.73	132.64	108.54	221.63
25	59.34	88.94	126.34	157.31	203.23	268.13
30	64.93	100.31	144.94	181.93	236.93	314.64
Length (μm) ↓			**16-nm Technology**			
10	137.01	184.89	244.78	291.87	374.78	461.67
15	160.69	225.32	313.58	384.67	493.20	652.78
20	181.31	265.58	378.45	481.56	621.28	817.88
25	204.89	314.77	455.21	574.34	748.55	994.82
30	222.45	358.11	526.89	666.13	876.60	1171.70

TABLE 2.13
Length-dependent resistance analysis of SWCNT bundle interconnects at different temperatures

Length (μm)↓	32-nm Technology node								
	Densely Packed			Moderately Packed			Sparsely Packed		
Temperature (K)→	**100**	**300**	**450**	**100**	**300**	**450**	**100**	**300**	**450**
10	**12.01**	**17.33**	**63.53**	**13.55**	**19.77**	**70.23**	**20.01**	**28.88**	**77.08**
20	**22.01**	**30.45**	**70.45**	**23.01**	**33.01**	**75.57**	**25.05**	**37.74**	**79.99**
30	**29.56**	**39.33**	**80.00**	**30.24**	**43.73**	**86.66**	**34.06**	**40.04**	**89.98**
Length↓	**16-nm Technology node**								
	Densely Packed			**Moderately Packed**			**Sparsely Packed**		
Temperature (K)→	**100**	**300**	**450**	**100**	**300**	**450**	**100**	**300**	**450**
10	**68.88**	**122.56**	**230.33**	**70.44**	**134.22**	**250.66**	**82.02**	**144.21**	**300.12**
20	**72.11**	**129.00**	**250.22**	**77.33**	**202.11**	**266.56**	**85.32**	**132.12**	**311.02**
30	**84.12**	**135.32**	**278.77**	**88.12**	**232.53**	**298.99**	**90.11**	**140.11**	**333.23**

of these metallic nanotubes within SWCNT bundle interconnects varies from process to process. The temperature-dependent resistance analysis in Table 2.13 is carried out for three different cases of carbon nanotube distribution (densely packed, moderately packed and sparsely packed) in the given SWCNT bundle with all CNTs of fixed diameter. The result obtained reveals that both the technology node (32 nm and 16 nm) value of resistance rises with an increase in spacing. The reason behind this is that as we move from densely packed, moderately packed and sparsely packed SWCNT bundles, the porosity, i.e. spacing, increases; now with the increase in spacing, the effective number of conducting channels decreases and henceforth the resistance increases.

Furthermore, the tabulated results point out that for the higher side of the temperature range the resistance of SWCNT is severely impacted by the effect of temperature fluctuations and hence need to be addressed for efficient performance of any ASIC. Moreover, on a temperature-dependent comparative analysis of Cu with SWCNT for global level interconnect length at 32 nm and 16 nm technology nodes, we infer that SWCNT bundle interconnects exhibit less resistance compared to copper interconnects for all temperature values under consideration. Since the electrical performances of nano interconnects are affected by temperature and size, which may seriously limit the current density and the reliability, hence it is mandatory to include the influence of temperature on variable interconnect lengths for deep sub-micron technology nodes to evaluate the accurate performance of on-chip nano interconnects.

2.6 SUMMARY

- The prime intention of this chapter is to discuss the evolution of interconnect networks with the advancement in process technology.
- The sections of this chapter explain the criteria for selecting the wire model.
- A detailed explanation on the limitations of conventional interconnects is given, highlighting the challenges of emerging technology.
- In the end, it examines the properties of state-of-the-art graphene (CNT and GNR) interconnects.
- Based on the properties, the electrical model is developed to quantify the impact of wire material on performance and power consumption.
- A comparative analysis of conventional and state-of-the-art interconnects is carried out for different technology nodes and at different interconnect levels (global, intermediate and local) based on length.
- A thermal stability analysis is also done to observe the effect of operating temperature fluctuation on the performance of ASIC.

EXERCISES

MULTIPLE-CHOICE QUESTIONS

Q1 Higher-level metal layers have __ thickness compared to lower-level metal layers.
a Larger
b Equal
c Smaller
d All of these

Q2 The thermal conductivity of a standard SWCNT along its length is ___ W/mK.
a 3,500

b 385
c 35,000
d 35

Q3 MWCNT possesses electrical superconductivity up to a temperature of _____.
a 12K
b 12°C
c 100K
d 100°C

Q4 Who prepared and explained nanotubes for the first time?
a Sumio Iijima
b Richard Smalley
c Eric Drexler
d Richard Feynmann

Q5 The compressive strength of a nanotube _____ its tensile strength.
a is less than
b is greater than
c is equal to
d none

Q6 The electrical conductivity of a nanotube is ____ times that of copper.
a 10
b 100
c 1,000
d 1/100

Q7 Carbon nanotubes, often called the strongest material, have which of the following properties?

i. High electrical and thermal conductivity
ii. Very high tensile strength
iii. Higher lifetime

Select the correct answer using the codes given below:
a i only
b i and ii only
c i and iii only
d i, ii and iii

Q8 Which of the following is NOT a type of carbon nanotube structure?
a Armchair
b Chiral
c Zigzag

d Helix

Q9 Which of the following is NOT a way to synthesize CNT?
a CVD
b Laser ablation
c Electric arc discharge
d Quantum dot synthesis

Q10 Vapor-phase epitaxy is based on which one of the following?
a CVD
b Diffusion
c Ion implantation
d PVD

Q11 The first talk about nano technology was given by ___________.
a Albert Einstein
b Newton
c Gordon E. Moore
d Richard Feynman

Q12 A unit of sheet resistance for a Cu conductor is ___________.
a ohm/square
b ohm
c ohm m
d ohm/m

Short-Answer Questions

Q1 Explain in brief what you understand by the terms "interconnect density" and "technology integration."

Q2 What is more than Moore's law?

Q3 Enlist the three types of interconnects on the basis of length of wire. Mention their features.

Q4 On the basis of hybridization, give reasons why CNT is more reactive than planar graphene.

Q5 What are the advantages of MWCNT over SWCNT?

Q6 What are the advantages of GNRs over CNTs?

Q7 What is skin depth and how is the resistance of Cu affected by it?

Long-Answer Questions

Q1 What are the different methodologies adopted for CNT synthesis?

Q2 Define the term *chirality*, explain its significance in the context of carbon nanotube modeling and also determine the expression for CNT diameter in terms of its chiral vector.

Q3 Differentiate between the following.
a Armchair and Zigzag CNT
b Top-contact and side-contact GNR

Q4 Explain the basic structure of GNR. What are the different types of GNR? Classify them as metallic or semiconducting.

Q5 What are the conventional interconnect and state-of-the-art interconnect materials? What are the limitations of conventional interconnect materials that caused them to be replaced by state-of-the-art interconnects?

Q6 What are CNTs? What are the different types of CNTs? Highlight the important properties of CNTs.

1. A	7. B
2. A	8. D
3. A	9. D
4. A	10. A
5. A	11. D
6. C	12. A

ANSWERS TO MULTIPLE-CHOICE QUESTIONS

REFERENCES

Aoki, Hideo and Mildred S. Dresselhaus, ed. *Physics of Graphene* Springer: Heidelberg, 2014.

Avouris Phaedon, Joerg Appenzeller, Richard Martel, Shalom J. Wind. "Carbon Nanotube Electronics". *Proceedings of IEEE* 91, no. 11 (2003): 1772–1784.

Banadaki, Yaser M., Safura Sharifi, Walter O. Craig III, Hsuan-Chao Hou. "Power and Delay Performance of Graphene-based Circuits Including Edge Roughness Effects." *American Journal of Engineering Research (AJER)* 5, no. 6 (2016): 266–277.

Barke, Erich. "Line-to-Ground Capacitance Calculation for VLSI: A Comparison." *IEEE Transactions on Computer-Aided Design of Integrated Circuits and Systems* 7, no. 2 (1988): 295–298.

Burke, Peter John. "Luttinger Liquid Theory as a Model of the Gigahertz Electrical Properties of Carbon Nanotubes." *IEEE Transactions on Nanotechnology* 1 (2002): 129–144

Burke, Peter John. "An RF Circuit Model for Carbon Nanotubes." *IEEE Trans*actions on *Nanotechnology* 2 (2003): 55–58.

Dokania Rajeev K. "Analysis of Challenges for On-Chip Optical Interconnects", *GLSVLSI* (2009).

Elgamel, Mohamed A., Magdy A. Bayoumi. "Interconnect Noise Analysis And Optimization in Deep Submicron Technology." *IEEE Circuits and System Magazine* 3, no. 4 (2003): 6–17.

Emtsev Konstantin V., Bostwick Aaron, Horn Karsten, Jobst Johannes, Kellogg Gary L., Ley Lothar, Jessica L. McChesney, Taisuke Ohta, Sergey A. Reshanov, Jonas Röhrl, Eli Rotenberg, Andreas K. Schmid, Daniel Waldmann, Heiko B. Weber, and Thomas Seyller. "Towards Wafer-Size Graphene Layers by Atmospheric Pressure Graphitization of Silicon Carbide." *Nature Materials* 8, no. 3 (2009): 203–207.

Fetter, Alexander L. "Electrodynamics of a Layered Electron Gas. I. Single Layer." *Annals of* 81 (1973): 367–393.

Fisher, Matthew P. A., and Leonid I. Glazman. "Transport in one Dimensional Luttinger Liquid." In Mesoscopic Electron Transport, edited by L. L., Sohan, L. P., Kouwenhoven, and G., Schoen, NATO ASI Series, 345 (1996): 331.

Goel, Ashok K. High-Speed VLSI Interconnections: Second Edition (2007): 1–407. 10.1002/9780470165973

Hamedani, Soheila Gharavi and Mohammad Hossein Moaiyeri. "Comparative Analysis of the Crosstalk Effects in Multilayer Graphene Nanoribbon and MWCNT Interconnects in Sub-10 nm Technologies." *IEEE Transactions on Electromagnetic Compatibility*, 62, no. 2, (April 2020): 561–570, doi: 10.1109/TEMC.2019.2903567.

Hamedani, Soheila Gharavi and Mohammad Hossein Moaiyeri. "Impacts of Process and Temperature Variations on the Crosstalk Effects in Sub-10 nm Multilayer Graphene Nanoribbon Interconnects." *IEEE Transactions on Device and Materials Reliability* 19, no. 4, (2019): 630–641, doi: 10.1109/TDMR.2019.2937789.

Han, Melinda Y., Barbaros Özyilmaz, Yuanbo Zhang and Philip Kim. "Energy Band-gap Engineering of Graphene Nanoribbons." *Physical Review Letters* 98, no. 20 (2007): 206805.

Iijima, Sumio. "Helical Microtubules of Graphitic Carbon." *Nature* 354, no. 6348 (November 1991): 56-58.

Imry, Yoseph and Rolf Landauer. "Conductance Viewed as Transmission." *Reviews of Modern Physics* 21 (1999): 5306–5312.

International Technology Roadmap for Semiconductors (ITRS). [Online]. Available: http://public.itrs.net, (2013).

Kamon Mattan, Michael J. Tsuk, and Jacob K. White. "FASTHENRY: A Multipole-Accelerated 3-D Inductance Extraction Program." *IEEE Transactions on Microwave Theory and Techniques*,42, no. 9, (Sep. 1994): 1750–1758.

Kaustav, Banerjee and Amit Mehrotra. "Inductance Aware Interconnect Scaling." In Proceedings of International Symposium on IEEE Quality Electronic Design, (2002).

Khursheed, Afreen and Kavita Khare. "Designing Dual-Chirality and Multi-Vt Repeaters for Performance Optimization of 32 nm Interconnects." *Circuit World* 46, no. 2 (2020): 71–83. doi: 10.1108/CW-06-2019-0060.

Koester, Steven J., Albert M, Young, Roy, Yu, Sampath, Purushothaman, Kuan-Neng, Chen, Douglas Charles, La Tulipe, N. Rana, Shi, Leathen, Wordeman, Matthew R., and Edmund J. Sprogis. "Wafer-Level 3D Integration Technology." *IBM Journal of Research and Development* 52, no. 6 (2008): 583–597.

Kronig R. De L. and William George Penney. "Quantum Mechanics of Electrons in Crystal Lattices." *Proceedings of the Royal Society (London) A* 130 (1931): 499–513.

Kumar, Vachan, Shaloo Rakheja and Azad Naeemi. "Performance and Energy-Per-Bit Modelling of Multilayer Graphene Nanoribbon Conductors," *IEEE Transactions on Electron Devices* 59, no. 10, (Oct 2012).

Li, Baozhen, Timothy D., Sullivan, Tom C., Lee and Dinesh, Badami. "Reliability Challenges For Copper Interconnects." *Microelectronics Reliability*, 44 no. 3 (2004): 365–380.

Li, Hong, Wen-Yan Yin, Kaustav Banerjee, Jun-Fa Mao. "Circuit Modeling and Performance Analysis of Multi-Walled Carbon Nanotube Interconnects." *IEEE Transactions on Electron Devices* 55, no. 6 (2008): 1328–1337.

Li, Jun, Qi Ye, Alan Cassell, Hou Tee Ng, Ramsey Stevens, Jie Han and M. Meyyappan. "Bottom-Up Approach For Carbon Nanotube Interconnects." *Applied Physics Letters* 82 (2003): 2491–2493.

Li, J., Q. Ye, A. Cassell, H.T. Ng, R. Stevens, J. Han and M. Meyyappan. "Bottom-Up Approach For Carbon Nanotube Interconnects." *Applied Physics Letters* 82, no.15b (2003): 2491–2493.

Lim, C. K., T., Devolder, and C. Chappert. "Domain Wall Displacement Induced by Subnanosecond Pulsed Current." *Applied Physics Letters* (2004).

Maffucci, Antonio, Giovanni, Miano, and Fabio, Villone. "A Transmission Line Model for Metallic Carbon Nanotube Interconnects." *International Journal of Circuit Theory and Applications* 36 (2008): 31–51.

Maffucci, A., G. Miano and F. Villone. "A New Circuit Model for Carbon Nanotube Interconnects With Diameter-Dependent Parameters." *IEEE Transactions on Nanotechnology* 8, no. 3 (2009): 345–354

Massoud, Yehia and Arthur, Nieuwoudt. "Modeling and Design Challenges and Solutions for Carbon Nanotube-based Interconnect in Future High Performance Integrated Circuits." *ACM Journal on Emerging Technologies in Computing Systems* 2 (2006): 155–196.

Mayadas, A.F. and M. Shatzkes. "Electrical-Resistivity Model for Polycrystalline Films: The Case of Arbitrary Reflection at External Surfaces." *Physical Review B* 1 (1970): 1382–1389.

Meindl, James D. "Beyond Moore's Law: The Interconnect Era." *Computer Science and Engineering* (2003): 20–24.

Miano, Giovanni and Fabio Villone. "An Integral Formulation for the Electrodynamics of Metallic Carbon Nanotubes Based on a Fluid Model." *IEEE Transactions on Antennas and Propagation* 54, no. 10 (2005): 2713–2724.

Murali, Raghunath, Kevin Brenner, Yinxiao Yang, Thomas Beck, and James D. Meindl. "Resistivity of Graphene Nanoribbon Interconnects." *IEEE Electron Device Letters* 30, no. 6 (Jun 2009): 611–613.

Naeemi, Azad, Chang, Sou-Chi, Sourav, Dutta, Chenyun, Pan, Sasikanth, Manipatruni, Dmitri, Nikonov, and Young, Ian. Spin-Based Interconnect Technology and Design *2016 IEEE International Interconnect Technology Conference / Advanced Metallization Conference (IITC/AMC)*, San Jose, CA, (2016): 1–64, doi: 10.1109/IITC-AMC.2016.7507735.

Naeemi, Azad and James D. Meindl. "Conductance Modeling for Graphene Nanoribbon (GNR) Interconnects." *IEEE Electron Device Letters* 28, no. 5 (May 2007): 428–431.

Naeemi, Azad and James D. Meindl. "Monolayer Metallic Nanotube Interconnects: Promising Candidates for Short Local Interconnects", *Electron Device Letters* 26, no. 8 (2005): 544–546.

Naeemi, Azad, Reza Sarvari, and James D. Meindl. "Performance Comparison Between Carbon Nanotube and Copper Interconnect for Gigascale Integration (GSI)." *IEEE Electron Device Letters* 26, no. 2(Feb 2005): 84–86.

Nakada, Kyoko, Mitsutaka, Fujita, Katsunori, Wakabayashi, and Kusakabe,Koichi. "Localized Electronic States on Graphite Edge." *Czechoslovak Journal of Physics* 46 (1996): 2429–2430. doi: 10.1007/BF02570201.

Ngo, Duc-The, Kotato, Ikeda, and Hiroyuki, Awano. "Direct Observation of Domain Wall Motion Induced by Low-Current Density in TbFeCo Wires." *Applied Physics Express* (2011).

Nihei, Mizuhisa, Kondo, Daiyu, Kawabata, Akio, Sato, Shintaro, Shioya, H., Sakaue, M., Iwai, Taisuke, Ohfuti, Mari, and Awano, Yuji. Low-Resistance Multi-walled Carbon Nanotube Vias With Parallel Channel Conduction of Inner Shells, *In Proceedings of the IEEE International Interconnect Technology Conference*, (2005): 234–236.

Neto, Antonio H. Castro, Francisco Guinea, Nuno M. R. Peres, K. S. Novoselov, and Andre K. Geim. "The electronic properties of graphene." *Reviews of Modern Physics* 81 (Jan 2009): 109–162.

Nieuwoudt, Arthur and Yehia Massoud. "Understanding the Impact of Inductance in Carbon Nanotube Bundles for VLSI Interconnect Using Scalable Modeling Techniques." *IEEE Transactions on Nanotechnology* 5, no. 6 (2006): 758–765.

Nishad, Atul K. and Rohit Sharma. "Analytical Time-domain Models For Performance Optimization of Multilayer GNR Interconnects." *IEEE Journal of Selected Topics in Quantum Electronics* 20, no. 1, (Jan./Feb 2014): 17–24.

Nishad, Atul K. and Rohit Sharma. "Self-Consistent Capacitance Model For Multilayer Graphene Nanoribbon Interconnects." *Micro Nano Letters*10, no. 8 (2015): 404–407.

Novoselov, K.S., A.K. Geim, S. Morozov, D. Jiang, Y. Zhang, S. Dubonos, et al. "Electric Field Effect in Atomically Thin Carbon Films." *Science* 306, no. 5696 (2004): 666–669.

Obraztsov, Alexander N. "Chemical Vapour Deposition: Making Graphene on a Large Scale." *Nature Nanotechnology* 4, no. 4 (2009): 212–213.

O'Connor, Ian and Frédéric, Gaffiot. "Advanced Research in On-Chip Optical Interconnects", in Piguet, C. (Ed.), *Lower Power Electronics and Design*, CRC Press, Boca Raton, FL. 2004.

Permission: http://www.itrs2.net/itrs-reports.html. 2017. https://www.intel.com/content/www/us/en/silicon-innovations/intel-14nm-technology.html. 2017.

Pepeljugoski,Petar, Jeffrey, Kash, Fuad, Doany, Daniel, Kuchta, Laurent, Schares, Clint, Schow, Marc, Taubenblatt, Offrein, Bert Jan, and Alan, Benner. "Low Power and High Density Optical Interconnects for Future Supercomputers." *Optical Fiber Communication (OFC)* 2010.

Pu, Shao-Ning, Wen-Yan Yin, Jun-Fa, Mao, Qing H. Liu. "Crosstalk Prediction of Single- and Double-Walled Carbon-nanotube (SWCNT/DWCNT) Bundle Interconnects."*IEEE Transactions on Electron Devices* 56, no. 4 (2009): 560–568.

Raychowdhury, Arijit and Kaushik Roy. "A Circuit Model for Carbon Nanotube Inteconnects: Comparative Study with Cu Interconnects for Scaled Technologies". ICCAD (Nov. 2004): 237–240.

Ruehli, Albert E. and Pierce A. Brennan. "Efficient Capacitance Calculations For Three-dimensional Multiconductor Systems." *IEEE Transactions on Microwave Theory and Techniques* MTT-21, no. 2, (Feb. 1973): 76–82.

Satio, Riichiro, Mitsutaka Fujita, G. Dresselhaus and M.S. Dresselhaus. "Electronic Structure of Graphene Tubules Based on C60." *Physical Review B* 46 (1992): 1804–1811.

Scarselli, Manuela, Paola Castrucci and Maurizio De Crescenzi. "Electronic and Optoelectronic Nano-Devices Based on Carbon Nanotubes." *Journals of Physics Condensed Matter* 24, no. 31 (2012): 313202-1–313202-36.

Sondheimer Ernest Helmut. "The Mean Free Path of Electrons in Metals." *Advances in Physics* 1, no. 1 (1952): 1–2.

Salahuddin, Sayeef, Mark Lundstrom and Supriyo Datta. "Transport Effects on Signal Propagation in Quantum Wires."*IEEE Transactions on Electron Devices* 52, no. 8 (2005): 1734–1742.

Srivastava, Ashok, Yao, Xu, and Ashwani K., Sharma. "Carbon Nanotubes For Next Generation Very Large Scale Integration Interconnects." *Journal of Nanophotonics* 4 (2010): 1–26.

Sarto, Maria Sabrina, Alessio Tamburrano. "Single Conductor Transmission-Line Model of Multiwall Carbon Nanotubes." *IEEE Transctions on Nanotechnology* 9, no. 1 (2010): 82–92.

Van Noorden, R. "Moving Towards a Graphene World." *Nature* 442, no. 7100 (2006): 228–229.

Varnava, Christiana. "Electronics in an Organic Package." *Nature Electronics* 3 (2020): 733, doi: 10.1038/s41928-020-00522-4.

Wilson, Linda. International Technology Roadmap for Semiconductors (ITRS) reports, 2006, http://www.itrs.net/reports.html.

Wong, Terence K.S. "Time Dependent Dielectric Breakdown in Copper Low-k Interconnects: Mechanisms and Reliability Models." *Materials (Basel)* 5, no. 9 (2012): 1602–1625. Published 2012 Sep 12. doi: 10.3390/ma5091602.

Xie, Liming, Hailiang Wang, Chuanhong Jin, Xinran Wang, Liying Jiao, Kazu Suenaga, and Hongjie, Dai. "Graphene Nanoribbons From Unzipped Carbon Nanotubes: Atomic Structures, Raman Spectroscopy, and Electrical Properties." *Journal of the American Chemical Society* 133, no. 27 (2011): 10394–10397

Xu, Yao, Ashok. Srivastava and Ashwani K. Sharma. "A Model of Multi-Walled Carbon Nanotube Interconnects. in Proceedings of the 52nd IEEE Midwest Symposium on Circuits and Systems, (2009): 987–990.

Xu, Chuan, Hong Li and Kaustav Banerjee. "Graphene Nano-Ribbon (GNR) Interconnects: A Genuine Contender or a Delusive Dream?." in Proceedings of IEEE International Electron Devices Meeting, San Francisco, CA, pp. 1–4, 15-17 Dec. 2008.

Yagmurcukardes, Mehmet, F. M. Peeters, Tuğrul Senger, Hasan Sahin. "Nanoribbons: From fundamentals to state-of-the-art applications." *Applied Physics Reviews* 3 (2016): 041302.

Yuan C. and Timothy N. Trick. "A Simple Formula for the Estimation of the Capacitance of Two-Dimensional Interconnects in VLSI Circuits." *IEEE Electron Device Letters* EDL-3, (Dec. 1982): 391–393.

3 Repeater Buffer Modeling

3.1 BACKGROUND

Aggrandizement in deep sub-micron technology although has ensued in the manufacturing of smaller and faster devices, but still the prospective of further improvements in performance has been restrained by the malignant effects caused by dimension scaling on interconnectivity, which dominates an IC layout. Such a great impact of interconnects on the system performance in the DSM era pushes the researchers to tackle this hurdle early in the design process (Rabaey and Pedram, 1996). It has been observed that scaling the interconnect line width results in an increase of its resistance, thereby raising the need of circuits with high drive capability for on-chip signaling.

The main objective while laying a connection between any two nodes on the chip is to provide a minimum signal transmission delay path. The optimal solutions to resolve this elementary snag of delay minimization without incurring the overhead cost in terms of power and area are wire sizing, buffer insertion or a combination of both. Widening of the interconnect line has an imperceptible impact on the delay (Shigyo, 2000). Additionally, the coupling between the wire worsens as the interconnect thickness increases and the feature size shrinks so as to keep the interconnect wire resistance low. Hence, out of various suggested techniques, the buffer insertion methodology is most practical and convenient and thus widely accepted.

In this approach, as revealed in Figure 3.1, buffers as repeaters are stowed at certain intervals along an interconnect wire length in order to boost the signal each time it is influenced by the line parasitic (Glasser and Dobberpuhl, 1985). While the signal propagates along the interconnecting wire, a repeater helps to regenerate sharper transition edges, thus increasing the bandwidth of wire. In addition to that, the regenerative nature of the driver also improves the signal slew rate and reduces the interconnect response time by quickly discharging the load capacitance.

In the subsequent sections of this chapter, the strategy of judicial placement of buffers as repeaters or signal amplifiers is discussed in detail along with the device modeling techniques of smart repeaters' circuits. Apart from this, Chapter 3 also gives an insightful explanation on the adoption of the concept of repeater insertion methodology in VLSI on-chip interconnects for the purpose of signal restoration and delay reduction. It highlights the features of novel technology used for designing buffers to overcome the limitations encountered by buffer insertion technology in terms of area and power consumption.

DOI: 10.1201/9781003104193-3

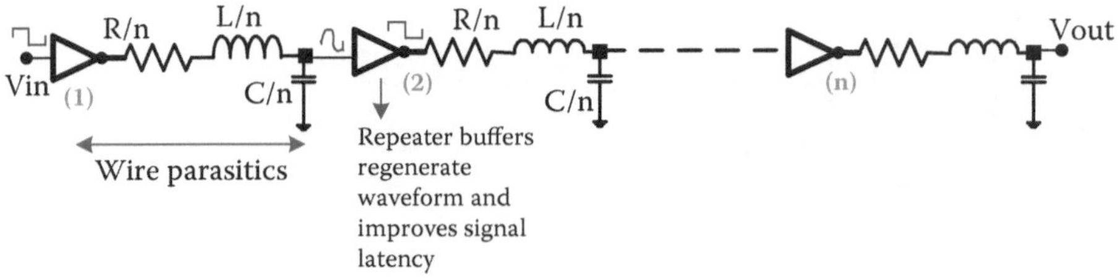

FIGURE 3.1 Interconnect with inserted buffers as repeaters in between.

3.2 NEED OF REPEATER INSERTION TECHNIQUE

The linear increase in interconnecting the wire length (l) causes a linear increase of both capacitance and resistance of wire simultaneously; thus making the Elmore (RC) delay of an un-buffered interconnect line as a quadratic function of wire length (l^2), despite the fact that the RC delay is not exactly an accurate estimate of signal latency of interconnects but is still widely used as a figure of merit. Hence to raise the operating speed of circuit; repeater insertion technique is adopted which transforms this square dependency of delay on length to linear one (Venkatesan, Jeffrey, and Meindl, 2003). The premier goal of this technique is to decimate lengthy interconnect wires into (N) tiny segments by interpolating buffers as repeaters in between every segment, as illustrated in Figure 3.2. Now the interconnect line has N segments with an overall delay of l^2/N.

Although interpolation of buffers seems to be an effective method for alleviating delay issue, the ramification of putting too many repeaters causes a switching power dissipation (Banerjee and Mehrotra, 2002). In addition to this, conventional buffers also fritter leakage power during the idle state. A peculiar trend has been observed that initially, over all, circuit delay decreases with the insertion of repeaters but after a requisite number of repeaters the delay increases (Alpert et al., 1999). This is because the repeater itself exhibits an additional switching delay or gate delay that contributes to an overall system delay. Hence, an optimum number and size of repeater buffers must be judicially decided upon, along with their strategically placed location to achieve better crosstalk noise immunity and minimum delay, without introducing an area and power penalty.

3.3 DESIGN CRITERIA OF REPEATER INSERTION

Even though the repeater insertion proves to be an effective scheme for mitigating overall wire delay, each repeater adds some gate delay. If the distance is too great between repeaters, the overall delay will be dominated by the long wires. If the distance is too small, the delay will be dominated by the large number of inverters. As usual, the best distance between repeaters is a compromise between these extremes. Moreover, it is apparent from the previous section that as we trail into the advanced ultra-deep sub-micron technology era, repeater interpolation is gradually becoming a massive trouble and bottleneck situation, due to their many numbers

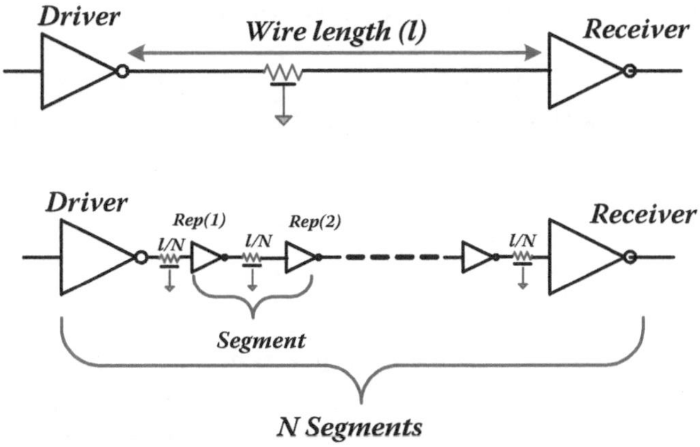

FIGURE 3.2 Repeaters are inserted by fragmenting a wire of length l into N segments.

and due to eventually becoming a major source of power dissipation. Thus, there is an imperative need to determine the optimal amount of repeaters by taking into account the individual repeater delay added to the overall repeater system and then finding the amount of repeaters inserted at which the increase in repeater delay exceeds the lower wire delay. This concept can be understood with the help of an expression for wire delay (t_d) with embedded repeater circuits with repeater delay (t_{rep}), given by Eq. 3.1.

$$t_d = \left[R_d c_w l + \frac{r_w c_w l^2}{2k} + (k - 1) t_{rep} \right] \tag{3.1}$$

If there is not any repeater buffer, i.e. k is equal to one, then in that case the wire delay (t_d) is the function of only the first two terms representing driver load with driver resistance (R_d)) and Elmore wire delay (with wire resistance and capacitance as r_w and c_w), respectively; and is independent of the third term introduced due to insertion of the buffer repeater. From the equation it is clear that as the wire line length l increases, its quadratic dependency causes (t_d) to rise abruptly; thus, the repeater interpolation technique may not be preferable for short interconnect lines but is effective for global interconnect lines.

Bakoglu and Meindl (1985) have introduced the formula for calculating the optimal number of repeaters denoted by (k_{opt}) and optimum size denoted by (h_{opt}) in the context of a wire propagation delay. Figure 3.3 depicts the various repeater insertion techniques.

The 50% propagation delay of the interconnect wire with k minimum size repeaters, each having an input capacitance, $\boldsymbol{C}_0$, and output resistance, $\boldsymbol{R}_0$ (Figure 3.3 (a)), is given by

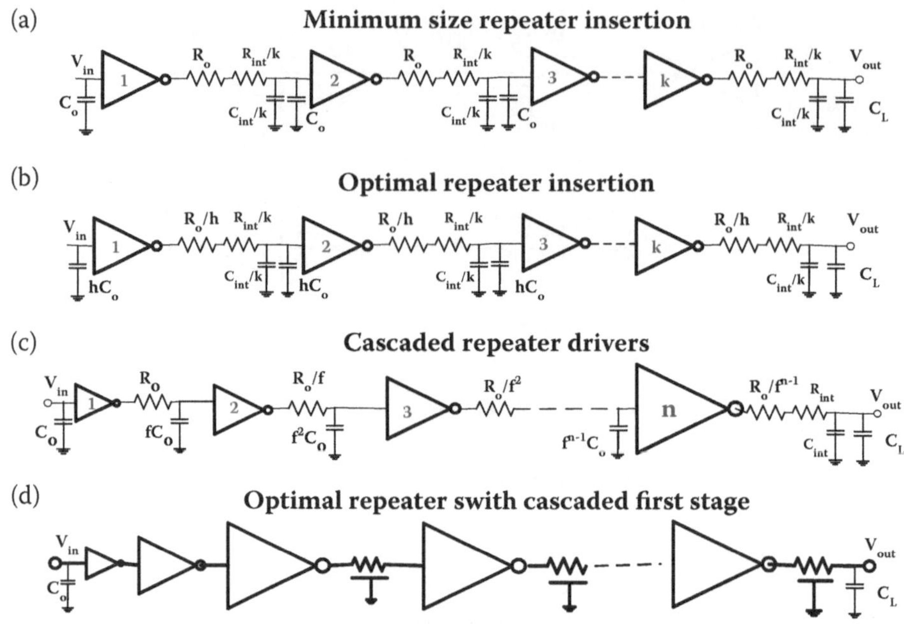

FIGURE 3.3 Types of repeater insertion techniques.

$$T_{50\%} = k\left[0.7R_0\left(\frac{C_{int}}{k} + C_0\right) + \frac{R_{int}}{k}\left(0.4\frac{C_{int}}{k} + 0.7C_0\right)\right] \tag{3.2}$$

To analyze the effect of variation in k on delay, differentiate $dT/dk = 0$

$$0.4\frac{R_{int}C_{int}}{k^2} = 0.7R_0C_0 \tag{3.3}$$

For achieving the minimum total delay value, the delay of segments connected by repeaters must be equal to that of a repeater. The optimum number of repeater values comes out to be

$$k = \sqrt{\frac{0.4R_{int}C_{int}}{0.7R_0C_0}} \tag{3.4}$$

The value of k should be at least two in order for this method to reduce the overall delay (Uno repeater is the driver). To implement this technique of minimum-size repeaters, it is advised to keep the RC constant of the wire, ($R_{int}C_{int}$), at least seven times the delay of a minimum-size repeater, (R_0C_0).

In addition to this, another technique for achieving further improvement in delay is by increasing the size of the repeaters (Figure 3.3(a)). The current driving capability of a repeater is directly proportional to W/L. When the W/L ratios of transistors are increased by a factor $\boldsymbol{h}$, the corresponding output resistance and input capacitance become R_0/h and $\boldsymbol{h}C_0$, respectively. The expression for delay will be

$$T_{50\%} = \boldsymbol{k}\left[0.7\frac{\boldsymbol{R}_0}{\boldsymbol{h}}\left(\frac{\boldsymbol{C}_{int}}{\boldsymbol{k}} + \boldsymbol{h}\boldsymbol{C}_0\right) + \frac{\boldsymbol{R}_{int}}{\boldsymbol{k}}\left(0.4\frac{\boldsymbol{C}_{int}}{\boldsymbol{k}} + 0.7\boldsymbol{h}\boldsymbol{C}_0\right)\right] \tag{3.5}$$

By evaluating $dT/dk = 0$ and $dT/dh = 0$, optimal values for $\boldsymbol{h}$ and $\boldsymbol{k}$ can be recapitulated as

$$\boldsymbol{k} = \sqrt{\frac{0.4\boldsymbol{R}_{int}\boldsymbol{C}_{int}}{0.7\boldsymbol{R}_0\boldsymbol{C}_0}} \tag{3.6}$$

$$\boldsymbol{h} = \sqrt{\frac{\boldsymbol{R}_0\boldsymbol{C}_{int}}{\boldsymbol{R}_{int}\boldsymbol{C}_0}} \tag{3.7}$$

And the resulting delay, which becomes relative to the geometric mean of the repeater and interconnect delay, can be expressed as

$$T_{50\%} = 2.5\sqrt{\boldsymbol{R}_0\boldsymbol{C}_0\boldsymbol{R}_{int}\boldsymbol{C}_{int}} \tag{3.8}$$

The two techniques mentioned are suitable for RC loads, whereas for driving large capacitive loads, the cascaded drivers technique is preferred. In this approach, in lieu of a single minimum-size driver, a successive string of drivers is used. These drivers are concatenated in order of their increase in size until the last buffer is large enough to drive the load, as depicted in (Figure 3.3 (c)). This method is efficient because if the load is driven by a large transistor, which in turn is driven by a small device, then the turn-on time of the large transistor dominates the overall delay. Thus, optimal delay is obtained with a trail of drivers concatenated with respect to their gradual increase in size. Hence, it can be concluded that this method optimizes the sum of the delay caused by charging the input capacitances of the cascaded drivers and the overall wire propagation delay, which can be expressed as

$$T_{50\%} = 0.7(n-1)fR_0C_0 + \left(\frac{0.7R_0}{f^{n-1}} + 0.4R_{int}\right)C_{int} + \left(\frac{0.7R_0}{f^{n-1}} + 0.7R_{int}\right)C_L \tag{3.9}$$

and by setting $\frac{dT}{dn}$ and $\frac{dT}{df}$ equal to zero, $f = e$ and $n = ln\left(\frac{C_{int}+C_L}{C_0}\right)$ we get the simplified expression of delay as

$$T_{50\%} = 0.7eR_0C_0 ln\left(\frac{C_{int} + C_L}{C_0}\right) + 0.4R_{int}C_{int} + 0.7R_{int}C_L \tag{3.10}$$

Figure 3.3 (d) depicts yet another technique in which the first stage of the optimal size repeaters must be a cascaded driver to minimize the input capacitance of the structure and to further optimize the overall propagation delay when it is driven by a minimum-size transistor.

Hence, it can be summarized that if repeaters are inserted, then the delay rises linearly with l_{int}. Preference is given to cascaded drivers if l_{int} is short as wire resistance is low for a small segment interconnect length, l_{int}, but for a longer wire segment length, l_{int}, delay rises expeditiously as cascaded drivers do not improve the $R_{int}C_{int}$ term. Moreover, in certain conditions, repeaters are not effective unless their sizes are optimized. Optimal size repeaters with a cascaded first stage obtain the shortest delay under all circumstances, at times reducing the overall delay by more than an order of magnitude.

In addition to this, splitting the long interconnecting wire line into shorter wire sections lowers the capacitive and inductive coupling between neighboring wires; consequently making it more immune to crosstalk noise (Saxena et al., 2004). However, in certain practical situations, it has been observed that the optimal position of a repeater buffer along the wire length could not be achieved because of physical space constraints. Even the change in buffer size and number of repeaters cannot compensate for variation in ideal physical placement. Hence, the upcoming research works are emphasizing more on arrangement of repeater buffers rather than on their number and size. Thus, to overcome the challenges encountered in routing and placement of repeater buffers integrated with on-chip interconnects, a buffer staggering technique is adopted. As depicted in Figure 3.4, in this technique repeater buffers are interleaved in such a way that the position of a buffer in any interconnect line is in middle of the two neighboring buffers that are placed in its adjacent lines.

By staggering the buffers, the worst-case capacitive coupling only persists for half the wire length and inductive coupling among the interconnect wires is also averaged. One more merit of this scheme that makes it the best-case repeater insertion technique among the remainder schemes depicted in Figure 3.3 (a–d) is that no additional area overhead cost is added besides assisting in delay reduction.

3.4 MODELING OF REPEATER BUFFER FOR ON-CHIP INTERCONNECTS

Buffer modeling is carried out bearing in mind that the minimum delay should be achieved without compromising the power dissipation or chip area.

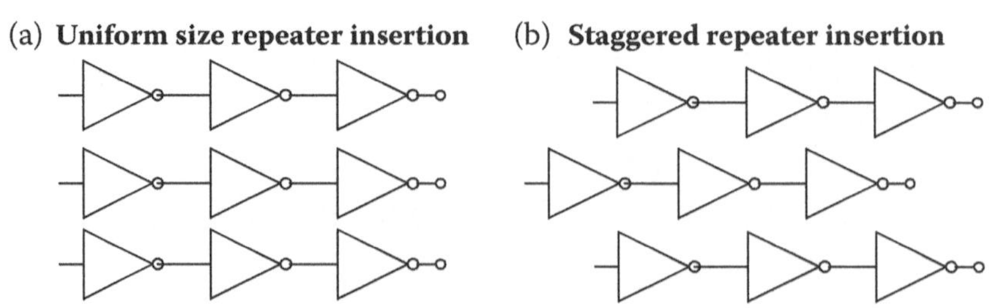

FIGURE 3.4 Staggering repeaters to reduce the worst-case delay and crosstalk noise.

3.4.1 Limitation of Conventional (CMOS) Buffer Fabrication Technology

Classical MOS-based fabrication technology has contributed to the chip-designing industries for years but with the advent of the nanometer device era, it has become quite hard to satisfy Robert Dennard's scaling law based on Moore's postulate. The main cause that restricts further scaling of MOS circuits is that in order to reduce the size of a MOS-based field effective transistor, the effective channel length has to be decreased. The repercussions of this feature reduction are drain-punch through, mobility degradation, sub-threshold slope degradation, dopant fluctuations, short channel effect and surge in leakage currents. In brief, it can be mentioned that slimming the gate oxide layer below 3 nm in order to suppress the gate leakage current causes fallible dielectric performance. Moreover, the falling trend in performance because of this striking effect is primarily due to short channel effects such as hot carrier effects, direct tunneling gate leakage, drain-induced barrier lowering and gate-induced drain leakage. In the same manner, some other physical constraints of Si-based CMOS is the stipulation of channel doping so as to subdue the short channel effects. The impact of channel doping is that it triggers the lowering of charge carrier mobility and also band to band tunneling (BTBT) effect as a consequence of impurity scattering and the transverse electric field. Furthermore, it results in I_{ON} value reduction because of the impact of capacitance in the depletion charge channel. Besides this, the activation of dopants results in a resistive effect and also lowers the drain current. Moreover, due to a tremendous rise in production cost and testing, one has to face the economic challenges. One more drawback of Si CMOS is the deficient lithography of generating CMOS devices below the wavelength of light. In order to overcome the hampered performance issues of Si devices due to the previous effects, the researchers are in an arduous quest for exploring novel material technologies in addition to alternate device techniques to integrate with present Si-based device technology. It has been experimentally investigated that the use of CNTs/GNRs in the channel region results in an innovative device called a CNTFET/GNRFET, which is generally faster and has less power dissipation compared to the bulk silicon transistors. In brief, it can be discussed that CNTs, initially discovered in 1991, are allotropes of graphene with a 2D honeycomb carbon lattice sheet, bundled in the form of a cylindrical tube. Governed by the chirality (the way they are rolled into tubes), CNTs show a metallic or semiconducting nature. The wrapping vector also determines the electrical characteristics of carbon nanotubes. As the diameter of CNTs is inversely proportional with the bandgap of a semiconductor, it is easier to control the threshold voltage of CNT-based FETs. The outstanding material properties of CNTs have a very high charge-carrying capacity, temperature variation stability and remarkable mechanical strength, thereby making it an excellent contender for future interconnect modeling material. Due to the lack of dangling bonds, it is unchallenging for carbon nanotubes to avoid a high *k* dielectric, which is a major issue in the case of GNRFETs. Graphene is a monolayer of graphite prearranged in a honeycomb hexagonal 2D lattice structure. It was introduced by the British and Russian physicists team Kostya and Andre Geim for which they were given the Nobel Prize in 2010. In simpler terms, unrolled CNTs can be visualized as graphene. This graphene is then introduced in the channel region of the field effective transistors to obtain graphene transistors. The outstanding properties are superior carrier mobility, better thermal conductivity and very fast switching due to a higher carrier velocity. Generally due to zero bandgap ($E_G = 0$), graphene demonstrates a metallic nature, but on patterning it to GNR of a few nanometers in width, then a small bandgap is introduced that changes its behavior to a semiconducting nature. As the energy bandgap holds an inverse proportionality with the width of ribbon, to improve the drive strength of GNRFET, multiple numbers of nanoribbons will be connected in parallel to reduce the effective resistance and increase the number of conducting channels. It can be understood clearly and in more detail from the next section the reason why carbon-based FETs are emerging as promising candidates for high-frequency and low-voltage post-Si electronics applications.

3.4.2 State-of-the-Art Emerging Buffer Fabrication Technology

Analogies are rarely perfect and sometimes can be misleading, but the comparison to a water pipe that controls the guided flow of water does provide a sense for the field-effect transistor (FET). A transistor is an electronic device that permits the guided flow of electrons with the help of a gate that controls the density of electron flow (this gate is analogous in logic to a valve controlling the amount of water). The field-effect transistor (FET) came into existence in 1960 and since then has formed the cornerstone of modern electronics by bringing a revolution to the field of computing, communications, automation and healthcare and fosters today's digital lifestyles. The previous two decades saw the development of many exciting new nanodevices; the coming decade will see burgeoning interest in nanocircuit research. Since it typically takes a decade before academic research comes to fruition in actual products, the time has already arrived to start exploring novel nanotechnology buffer circuits. The rising fashion of Moore's law has stretched to the horizon, and now the researchers are finding the potential prospects in replacing the Si-based CMOS technology with carbon material technology. By just moving up by one block of the chart of the periodic table, the carbon material that is there has quite analogous traits like Si. Both carbon and silicon hold the same count of valence electrons in their outermost valence shell, but with variation in their Coulomb interactions and henceforth have differences in size of electronic wave functions. Thus, researchers are extensively promoting carbon-based devices as a prospective alternate for next-generation integrated circuits because of its extraordinary crystal structures and allotropes. The basic building blocks of all these allotropes are C–C atoms systematically arranged in a 2D honeycomb hexagonal lattice framework. Some of the potential carbon-based substitutes are nanowire (NW) transistors, quantum-dot cellular automata, carbon nanotube-based field effective transistor and graphene nanoribbon-based field-effect transistors. As no single nanotechnology is expected to dominate all market segments, this section of the book presents a review of the evolution of some of the state-of-the-art buffers that have been developed so far using various nanotechnologies.

3.4.2 (A) State-of-the-Art Emerging (Nanowire Transistors) Buffer Fabrication Technology

Nanowire (NW)-type field-effect transistor (NWFET)-based repeater buffers are also termed surround gate FETs and have a fine nanowire channel in between the source and drain regions for charge transportation. Recently, NW-type FETs have drawn notable attention in the emerging nanoscale regime and are taken as a suitable contender for persistently scaled CMOS due to their non-planar geometry. NWFETs are said to provide exceptionally notable electrostatic control over the conduction channel as juxtapose to other existing planar structures. The upcoming research in it stems from its inexpensive and highly economical fabrication process, higher-yield electronic properties, higher mobility of carriers and considerably better scalability as the nanowire diameter can be controlled down up to a range of 10 nm. As a consequence of a smaller radius, the inversion charge reverts from surface inversion to bulk inversion mainly because of quantum confinement. Apart from this, there is also a notable rise in the bandgap with a decrease in nanostructures. This effect of quantum confinements makes the modeling of nanowire FETs more trivial. NWFETs show a remarkable surface-to-volume ratio that makes them appropriate for sensing and better selectivity. The common applications of NWFETs, besides their use as repeater buffers for nano interconnects, are in designing buffers for biosensor applications (for biosensing different biomolecules like urea, glucose and cholesterol, which helps in preliminary screening of the diseases). Nanowire transistors also used for environmental and other industrial applications. FETs based on NW technology are likely to provide better amplifications of charge and also maintain a high SNR (signal-to-noise) ratio compared to other techniques of detection and sensing. Although still in their embroyic stage, NWFETs are analytically studied for on-chip nano interconnects applications.

3.4.2 (B) State-of-the-Art Emerging (Quantum-Dot Cellular Automata) Fabrication Technology

QCA (quantum-dot cellular automata) consists of a uniform grid of cells in discrete dynamical systems. At a specific instance of time, each cell can be in just one of a finite number of states. As the time proceeds, the state of each cell within the grid is found by a rule of transformation that is directed by the neighborhood of the cell with factors in its previous state and the states of the next adjacent cells. In general, the heat generated in the switching of the cycle can be no longer removed, thereby causing a notable mitigation on the speed of the device with the increase in count of transistors embedded on a single chip. An array of coupled quantum dots to realize Boolean logic is considered in order to solve the QCA. The pros of QCA technology include a better packing density because of the tiny size of dots as well as its less complicated interconnection and remarkably low energy delay product. With the advent of graphene-based quantum dots commonly termed graphene quantum dots (GQDs), there is a huge scope in its implementation because they are flat-surfaced zero-dimensional nanocarbons. The GQDs besides on-chip interconnects find uses in areas of energy storage like photovolatics, supercapacitors, LEDs, rechargeable batteries and photodetectors, whereas biomedical applications of GCDs include electrochemical sensing, photoluminescence sensing and also cellular imaging.

3.4.2 (c) State-of-the-Art Emerging (CNTFET) Buffer Fabrication Technology

In order to keep pace with Moore's Law, the past decades have witnessed a spectacular scaling of transistor elements, such that the number of transistors on an integrated circuit are doubling approximately every two years (Lundstrom, 2003;Wong, 2005;Chau et al., 2003). Integration and innovation of novel materials, such as high-k gate dielectrics, various metals, silicides and nitrides, are instrumental for this evolutionary path of CMOS device scaling. Without being affected by these innovations, the active channel material has predominantly remained the same to date, mainly due to the scalability and manufacturability of the Si technology. Ever since, the gate length of MOSFET has entered the deep sub-micron region, directly tunneling between source (S) and drain (D), and severe short channel effects raised a fundamental challenge in continued scaling of Si devices. As a result, a lot of efforts have been undertaken by many academic and industrial researchers for integrating new semiconductors as the channel material so as to enable (i) higher mobility and thereby resulting in efficient transport of carriers and (ii) improved electrostatics at the nanoscale (i.e. non-planar channel materials) (Datta et al., 2005 Kim and del Alamo, 2007).

As one of the promising new devices, carbon allotropes avoid most of the fundamental limitations for traditional silicon devices. All the carbon atoms in them are bonded to each other with sp^2 hybridization and there is no dangling bond that enables the integration with high-k dielectric materials. In the following section, we will we focus on CNTs and introduce graphene nanoribbons.

Carbon Nanotube

Much of the discussion on the physics of carbon nanotubes was already completed in Chapter 2, where it was explained that carbon nanotubes are one-dimensional (1D) graphene sheets rolled into a tubular form. Their electronic properties are governed by the tube diameter and wrapping angle determined by the chiral vector, which is characterized by the indices (n,m) of the graphene. In the previous chapter, much emphasis was given on assessing the potential use of CNTs as interconnects. In the following section, some insight into the carbon nanotube as a material for realization of a field-effect transistor will be presented.

Carbon Nanotube Field-Effect Transistor

The initial reports of operational room-temperature of a carbon nanotube field-effect transistor were obtained from IBM research team and the Delft University of Technology (Tans et al., 1998). Figure 3.5 shows the basic structures of CNTFETs. The CNT channel region is undoped, while the other regions forming the source and drain are heavily doped. The center-to-center distance between two adjacent nanotubes is called the pitch. Unlike Si-MOSFETs, both p- and n-type CNTFETs with

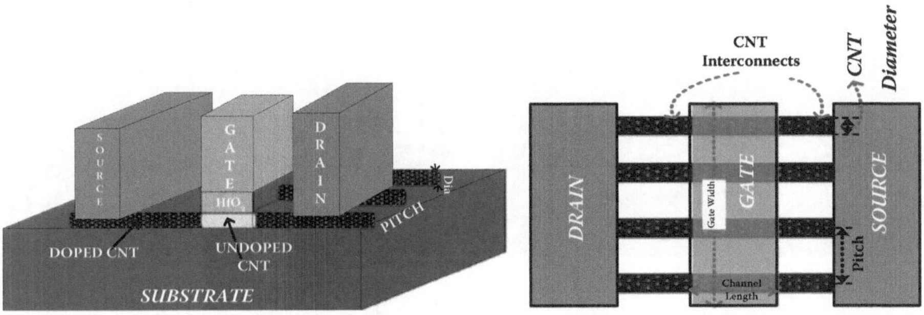

FIGURE 3.5 Basic structures of CNTFETs.

congruent geometry have the same mobility ($\mu_p = \mu_n$) and drive capability. The current flow can be increased by adding more CNTs or choosing a suitable diameter and chirality vector.

The three main key features that made the carbon nanotube field-effect transistor the most promising alternate technology to extend or complement traditional silicon technology are as follows:

- Firstly, the principle of operation and the device structure are analogous to existing CMOS devices, which makes it feasible for researchers to reuse the established CMOS design infrastructure.
- Secondly, it is convenient to use the existing CMOS fabrication process techniques.
- Thirdly, the most important reason is that CNFET has the best experimentally demonstrated device current carrying ability to date (Wong et al., 2003); the strong covalent bonding gives the CNTs high mechanical and thermal stability besides providing resistance to electromigration. The tube diameter is controlled by its chemistry and not by the standard conventional fabrication process. Furthermore, CNTs due to their ease of integration with high-*k* dielectric material, quasi-ballistic transport, large transconductance, higher sub-threshold slope and shorter intrinsic delay offers a great potential for nano ICs.

CNTFET Construction and Device Geometry

CNT field-effect transistors can have numerous geometrical configurations. Figure 3.6 depicts some of the widely used CNTFET device geometries. All four structures have similar attributes in terms of device geometry. The semiconducting nanotube acts as the channel for current transport, the source and drain form metallic contacts, a gate electrode is either in the form of a top-gate or a back-gate and an oxide or high-*k* dielectric layer provides insulation to the gate from the CNT.

The structure of a CNTFET is similar to the structure of a Si MOSFET, except for the use of CNT as the channel between two electrodes, which work as the source and drain of the transistor. The structure is built on top of an insulating layer and highly doped substrate. Figure 3.6 exhibits four common CNTFET device geometries. Figure 3.6(a) illustrates a back-gated CNTFET with the heavily doped (p++) Si

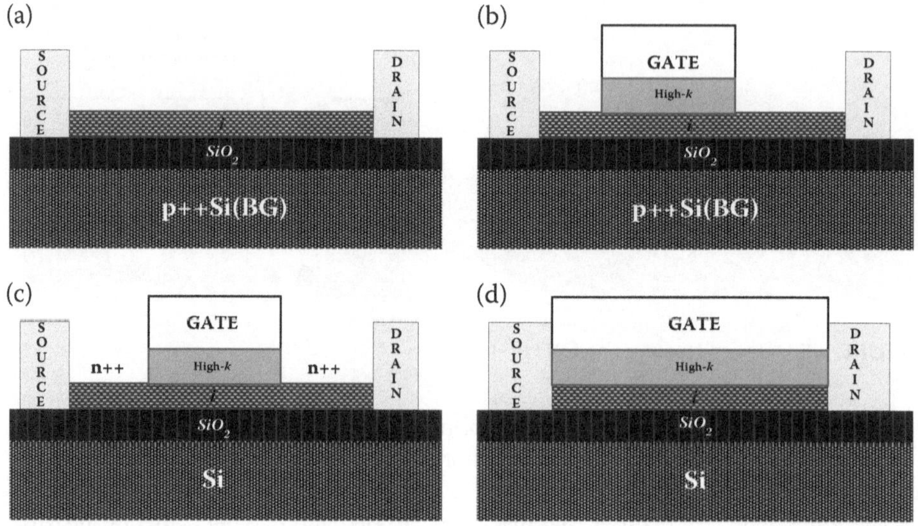

FIGURE 3.6 CNTFET device geometries.

substrate serving as the gate and the intrinsic (*i*) semiconducting CNT. Figure 3.6(b) shows CNTFET with both a top-gate and a back-gate; the top-gate controls the channel and the back-gate is used to electrostatically dope the CNT extension regions to achieve a low-resistance path to the source/drain contacts. Figure 3.6(c-d) shows a CNTFET with a chemically doped nanotube extension region and self-aligned CNTFET, respectively. A top-gate is often preferred over a back-gate to provide localized and greater control of the nanotube channel, as well as to afford individual gate control in a multi-transistor circuit offering high reliability, low voltage and high frequency operation.

CNTFET Operation and Working

State-of-the-art non-Si devices such as the carbon nanotube FETs (CNFETs) functions with entirely different device physics by way of a quasiballistic charge transport in the CNT-based channel.

The operating mechanism of a carbon nanotube field-effect transistor (CNFET) is analogous to that of traditional silicon devices. The CNT transistors can be made to turn on or off electrostatically by means of the gate. The quasi-1D device structure offers better gate electrostatic control over the channel region as compared to the 3D device (such as bulk CMOS) and 2D device (such as fully depleted SOI) structures.

CNT field-effect transistors on the basis of their principle of operation can be bifurcated as Schottky Barrier-type field-effect transistor (SB-CNTFET) and MOS-like FET (MOS-CNTFET). The working mechanism of these two types can be best explained with the help of energy band diagram shown in Figure 3.7.

As depicted in Figure 3.7(a), the conductivity of SB-CNTFET is governed by the majority charge carriers tunneling via the Schottky Barriers at the metal-nanotube

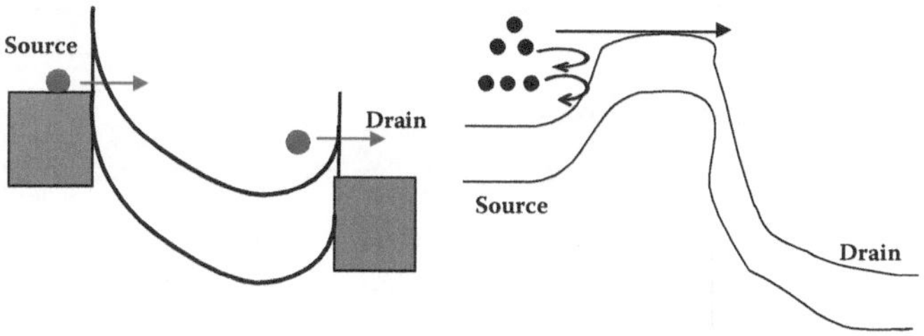

FIGURE 3.7 The energy band diagram for (a) SB-CNFET and (b) MOSFET-like CNFET.

end contacts. Here, the transistor ON-current governing the device performance is premeditated by the contact resistance instead of the channel conductance because of the tunneling barriers present at source and drain contacts. These Schottky Barriers appear at either one or both of the sources and drain contacts because of Fermi-level alignment occurring at the metal-semiconductor interface. The gate can electrostatically modulate the conductivity by altering the heights and widths of the Schottky Barriers. The capacity to direct the polarity of the nanotube FETs by just applying an appropriate metal contact while maintaining the nanotubes chemically intrinsic provides a unique path for novel device engineering. SB-CNTFET exhibits ambipolar transport behavior. The work function induced barriers at the end contacts can be made to improve the transport of charge carriers (electron/hole). Thus, both the device polarity (n CNTFET/p CNTFET) and the device bias point can be adjusted by choosing the appropriate work function of source/drain contacts, whereas MOS-CNTFET shows unipolar characteristics by suppressing either electron (p CNTFET) or hole (n CNTFET) transport with a heavily doped source/drain. As illustrated in Figure 3.7(b), the non-tunneling potential barrier is present in the channel region and, henceforth, the conductivity is thus modulated by the gate-source bias.

Research studies (Javey, 2004; Akbari Eshkalak, 2017; Javey et al., 2004) explain that up to a certain extent, good dc current can be achieved by SB-CNTFET with the self-aligned structure but its ac performance still seems pitiable due to proximity of the gate electrode to the metallic source/drain contact. Furthermore, the ambipolar nature of the SB-CNFET also makes it undesirable for complementary logic design. Considering both the fabrication feasibility (Javey et al., 2005) and superior device performance of the *MOS-CNTFET* with respect to the *SB-CNTFET*, we choose to focus on *MOS-CNTFET* in this book for studying the device operation, CNTFET on/off current in terms of device parameters and geometry.

Current Transport in CNTFET

Irrespective of the device geometry shown in Figure 3.6, the current transport CNT field-effect transistors can be set by the nanotube length juxtapose with their mean free path, l_m, and by the type of contact the nanotube makes with the source or drain metals. The charge transport regimes divided into four types, as tabulated in Table 3.1.

TABLE 3.1
Types of charge transport regimes

Metal-CNT contact	$L < l_m$	$L > l_m$
Ohmic	Ohmic-contact ballistic CNTFET	Ohmic-contact diffusive CNTFET
Schottky barrier	Schottky barrier ballistic CNTFET	Schottky barrier diffusive CNTFET

Among these four, CNTFET with a ballistic ohmic contact is proffered as it offers the smoothest flow of charge. Furthermore, we know that technically absence of electron scattering is termed ballistic transport. Due to this added benefit, it is much simpler to do the mathematical analysis.

Analysis of 1D Ballistic Carbon Nanotube Field-Effect Transistor

Figure 3.8 shows the summarized ballistic nanotransistors model. The potential on the gate electrode V_{GS} induces the majority charge carriers in the CNTFET channel, also modulates the crown of the energy band between the source and the drain. If the barrier in between the source and drain is lowered, then the current flows between the source and drain. Together with the fact that electrons within the carbon nanotube are confined in the graphene atomic plane and the motion of electrons in the CNT is strictly restricted because of the quasi-1D structure of the carbon nanotube. Now, since electrons may move freely along the axial direction of tube, all wide-angle scatterings are prohibited except the possibility of forward and backward scattering due to electron-phonon interactions.

In analyzing the ballistic CNTFET, all the scattering mechanisms can be neglected. For the reason that the current tends to stay constant throughout the channel, it can be computed at the crest of the energy barrier corresponding to the beginning of the channel. The salient feature of the ballistic method of transport is that at the crest of the barrier, electrons entering from the source satiate the $+k$ states, whereas the electrons approaching from the drain occupy the $-k$ states. μ_s and μ_d denote the Fermi Level at the source and drain, respectively. Subjected to the chirality vector and diameter of nanotubes, the periodic boundary conditions impinge restrictions on the availability of states (Avouris et al., 2003), thereby resulting in a discrete set of energy sub-band structures.

As mentioned previously, the gate potential controls the crest of the energy barrier and lowers the channel potential by V_{CNT}, thereby resulting in an accrual of charge into the CNT channel, Q_{CNT}. Accumulation of the charge Q_{CNT} causes a voltage drop, V_{GS}–V_{CNT}, across the high-k insulator, resulting in a lower energy band by V_{CNT}. But this model is not compatible with Hewlett Simulation Program with Integrated Circuit (HSPICE); hence, the proposed model by Raychowdhry and Kaushik Roy is considered in this manuscript (Raychowdhury et al., 2004). This model explains the computation of V_{CNT} through the fitting parameters of the channel charge, Q_{CNT}

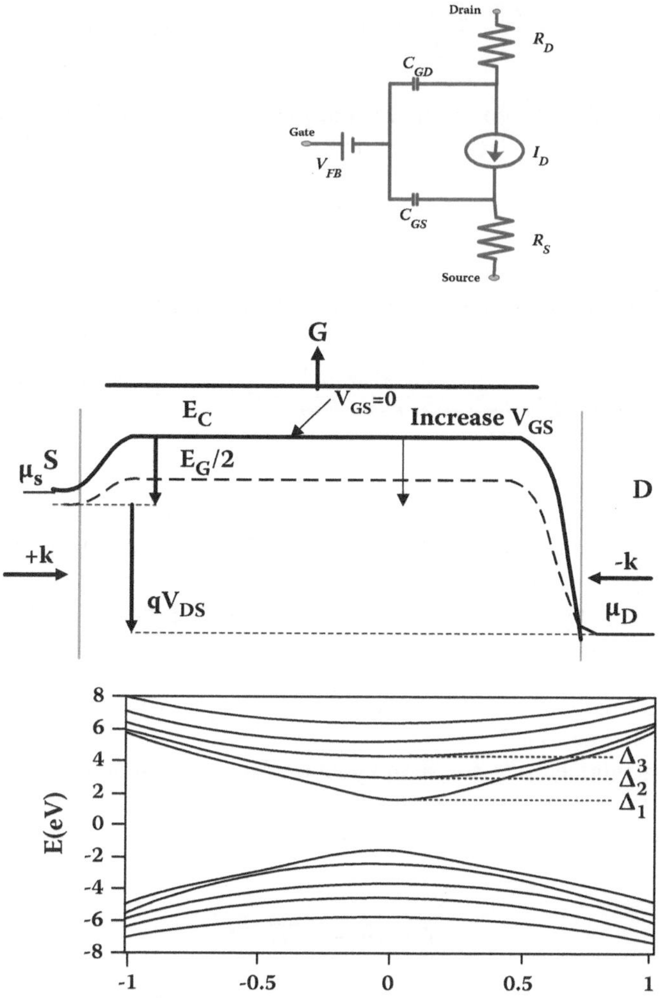

FIGURE 3.8 HSPICE simulation compatible circuit model for ballistic 1D-CNTFET and its energy band diagram.

$$V_{CNT} = V_{GS} - \frac{Q_{CNT}}{C_{ox}} \tag{3.11}$$

where C_{ox} is the gate-oxide capacitance, and Q_{CNT} is a function of number of carriers in the channel (n_{CNT}), which is the sum of the energy sub-band contributations individually by source and drain and can be computed using the mentioned equations:

$$n_{CNT} = \int_{E_c}^{+\infty} \frac{D(E)}{2} [f(E - \mu_s) + f(E - \mu_D)].\, dE \tag{3.12}$$

here, E_c is the basal of the conduction band, $f(E)$ is the probability that a state with energy E is occupied and $D(E)$ is the density of states (DOS) of CNT.

$$n_{CNT} = n_S + n_D \tag{3.13}$$

Using the Landauer formula, which explains the ballistic transport with ideal contacts, the equation for the drain-current can be obtained as

$$I_D = \frac{4ek_BT}{h} \sum_{p=1}^{+\infty} \left[ln\left(1 + \exp\frac{-\Delta_P + V_{CNT}}{k_BT}\right) - ln\left(1 + \exp\frac{-V_{DS} - \Delta_P + V_{CNT}}{k_BT}\right)\right] \tag{3.14}$$

where Δ_P denotes the pth energy sub-band, e denotes charge of an electron, k_B represents the Boltzmann constant, h represents the Planck constant and T denotes the ambient temperature.

Much of the discussion on CNT structure from the previous chapters' content will be instrumental to the readers for understanding basic device physics concepts involved in obtaining an electrical simulation compatible circuit model for CNTFET, with the lion's share of the content courtesy of Chapter 2. A brief review is present in this section.

It is aforementioned that a carbon nanotube is a rolled graphene sheet with the unit cell determined by a lattice vector $\vec{C}$ and a primitive translation vector $\vec{T}$. The wrapping geometry is defined by indices n *and* m, which specify the CNT diameter and chirality

$$\vec{C} = n\vec{a_1} + m\,\vec{a_2} \tag{3.15}$$

The dispersion relation for graphene, acquired from the Slater-Koster tight binding scheme, taking into consideration only the π orbital is expressed as

$$E_{g2D}(k_x, k_y) = \pm V_\pi \left\{1 + 4\cos\frac{\sqrt{3}\;k_xa}{2}\cos\left(\frac{k_ya}{2}\right) + 4cos^2\left(\frac{k_ya}{2}\right)\right\}^{\frac{1}{2}} \tag{3.16}$$

where k_x and k_y are wave vectors, V_π is the transfer integral (or the nearest-neighbor parameters). In order to make it compatible with the quasi one-dimensional CNT, the coordinate (k_x, k_y) is converted into (k_t, m). The wave vector k_t is in the direction of transport and m is the quantization number in the circumferential direction. Henceforth, the dispersion of CNT can be expressed as (Saito et al., 1998):

$$E_{CNT}(k_t, m) = \pm V_\pi \{1 + 4 \cos t_1 \cos t_2 + 4cos^2(t_2)\}^{\frac{1}{2}} \tag{3.17}$$

$$t_1 = \frac{\sqrt{3}a}{4}\frac{n_2 - n_1}{n}k_t + \frac{\pi}{2}\frac{n_1 + 3n_2}{n^2}m$$

$$t_2 = \frac{\sqrt{3}a}{4}\frac{n_2 + n_1}{n}k_t + \frac{\pi}{2}\frac{3n_1 - n_2}{n^2}m$$

$$n^2 = n_1^2 + n_2^2 + n_1 n_2$$

$$-\frac{\pi.\, d_R}{\sqrt{3}a.\, n} < k_t < +\frac{\pi.\, d_R}{\sqrt{3}a.\, n}$$

$$m = 0: \frac{2n}{d_R} - 1$$

$$d_R = \gcd(2n_1 + n_2\,,\ 2n_2 + n_1)$$

The density of states (DOS) is determined as

$$g(E) = \frac{2}{\pi}\sum\nolimits_m \int \left|\frac{\partial E_{CNT}}{\partial k_t}\right|^{-1} dE_{CNT} \tag{3.18}$$

In addition to this, the external potential that controls the charge and current should be reflected in the transistor theory so as to make it relevant with respect to the Ballistic transport in CNTs. Predominantly, the potential on the gate electrode gears the energy barrier for current flow between the source and drain electrodes, whereas the voltages of the source and drain correspondingly gear the source and drain Fermi levels. From the (I–V) characteristic graph (Akinwande et al., 2008), the threshold voltage can be determined as the voltage at which the gradient of the transconductance is a maximum. Analytically, it is determined by the following equation:

$$V_t \approx 0.5\left[\frac{E_g}{e}\right] = \frac{\sqrt{3}}{3}\left[\frac{aV_\pi}{eD_{CNT}}\right] \cong \frac{0.43}{D_{CNT}} \tag{3.19}$$

where a ($\cong$0.249 nm) is the carbon-carbon atomic distance, V_π = 3.033 eV pie to pie bond energy in tight bonding model and D_{CNT} is the diameter of the carbon nanotube and is determined as

$$D_{CNT} = a\sqrt{\frac{n_1^2 + n_2^2 + n_1 n_2}{\pi}} \cong 0.783\sqrt{n_1^2 + n_2^2 + n_1 n_2} \tag{3.20}$$

From Eqs. 3.19 and 3.20, it is concluded that V_t of CNTFET can be altered by changing the diameter value governed by the chirality vectors (n_1, n_2). The chiral angle (θ), which determines the direction of the chiral vector, is mathematically expressed as

$$\cos \theta = \frac{\frac{(n_1 + n_2)}{2}}{\sqrt{n_1^2 + n_2^2 + n_1 n_2}} \tag{3.21}$$

Changes in value of cos θ and D_{CNT} cause a variation in properties of CNTFET. From Table 3.2 it is clear that as the value of the diameter and chiral vector increases, then the value of V_t decreases, as the diameter is inversely proportional to V_t. Using the large diameter value increases the transconductance and reduces the threshold voltage consequential of which the switching speed increases as the device approaches saturation faster.

3.4.2 (D) State-of-the-Art Emerging (GNRFET) Buffer Fabrication Technology

Graphene Nanoribbon

Tantamount to CNTs, the graphene nanoribbons also exists in a number of forms. Single-layer graphene is obtained from one atomic-thick film of carbon, resulting in a pure 2D material. Unlike CNTs, the graphene sheets are not rolled and joint back to itself in order to form tubes; hence, the edges of carbon sheets are left free to

TABLE 3.2
Threshold voltages of CNTFET for different chirality vectors

Chirality	Diameter	$V_t(V)$
(4,0)	0.313	1.392
(7,0)	0.548	0.795
(10,0)	0.783	0.556
(13,0)	1.018	0.428
(16,0)	1.251	0.348
(19,0)	1.486	0.293
(22,0)	1.720	0.253
(25,0)	1.955	0.223

FIGURE 3.9 Graphene lattice structure.

bond with other atoms. These unconnected edges are not stable and usually passivated by absorbents like hydrogen.

Thus, it can be stated that graphene is a closely packed sheet of carbon-carbon atoms arranged in a 2D hexagonal framework. These sheets are etched into a 1D narrow stripe and are termed Graphene NanoRibbons (GNR). Smaller the width of nanoribbons, the more effect the edge structure has on its properties. The crystallographic orientation of the edges is a vital parameter. The graphene lattice structure in Figure 3.9 is based on the orientation of carbon atoms at the edge of the graphene sheet; GNRs are grouped into two types: zigzag GNRs (ZGNRs) and armchair GNRs (AGNRs). Normally AGNRs are semiconducting, whereas ZGNRs show metallic properties. In the case of armchair graphene nanoribbons, the bandgap defers to a periodic pattern related to N (dimer lines). If $N = 3p$ and $N = 3p + 1$, then the bandgap is finite and the nanoribbons of graphene exhibit semiconducting traits. If $N = 3p + 2$, then the bandgap is very minute, which results in nanoribbons of graphene to exhibit metallic traits (Son et al., 2006). With an increase in value of N, a falling trend in bandgaps is observed. Moreover, for armchair graphene nanoribbons, the bandgaps hold an inverse proportionality with the width of nanoribbons. In the case of zigzag graphene nanoribbons, the metallic characteristics are displayed if the edges are pristine, notwithstanding the fact that the bandgap might be opened for ZAGNRs having rough edges or those passivated by hydrogen atoms.

Armchair Graphene Nanoribbon

As per the width configuration of the nanoribbons, the armchair is classified as metallic or semiconducting; however, from analytical calculations obtained from discrete Fourier transform, armchair GNRs show semiconducting nature, with the width inversely proportional to its energy bandgap. The scanning tunneling microscopy technique is used to control the edge orientation of graphene nanoribbons.

Zigzag Graphene Nanoribbon

GNRs of the zigzag type are metallic with spin-polarized edges, whose gap can be open wide with antiferromagnetic coupling at the gap. Their behavior is directed by its edge state wave function, mainly governed by spin polarization due to its exchange interaction. The edge of zigzag GNR obtains the localized state of the edge with unbounded orbitals near the Femi level of energy. Due to this variation in electronic and optical properties, they experience virtue of quantization. From the tight binding theory, it can be stated that zigzag GNRs exhibit metallic traits, whereas the armchair shows semiconducting traits.

Process Technology Used for Fabrication of Graphene

- *Mechanical Cleavage*

The technique of mechanical cleavage helps in patterning graphite into monolayer atomic structures. Prior to the scotch tape technique, it is impossible to obtain one atomic-thick sheet of graphene. In order to get the samples of peeled graphene, the microchemical cleavage of bulk graphite is considered by employing adhesive tape to peel a layer of deeply oriented pyrolytic graphite. Once the peeling step is complete, then deposition is done on oxidized silicon for the substrate. The mechanical cleavage method is applied for pedagogical analysis of fundamental properties such as optical memories, temperature resistivity, ballistic transport and carrier mobility. Even though the mechanical cleavage technique produces satisfactory quality of graphene layers, it is not readily preferred for practical applications.

- *Chemical Exfoliation*

The chemical exfoliation technique has three stages: intercalation, reintercalation and sonication. In the intercalation stage, the graphite that is potentially expandable is made to be heated in nitric and sulphuric acids. The obtained exfoliated particles by this are in the naïve stage and multiple layers thick. In the succeeding step, reintercalation is carried out by doing an oleum treatment with tetra butyl alcohol. Finally, the sonication is carried out by sonicating graphite powder in N-methyl pyrrolidone.

- *Oxidation and Reduction*

The relatively efficient method to obtain a better yield of graphene is to initially synthesize graphite oxide and then exfoliating it into monolayers. In the next step, removal of the oxygen group is carried out with the help of the reduction reaction. Every graphene-oxidized flake has a great number of repelling negative charges. The consequence of oxygen treatment is the increase in interlayer spacing from 0.34 nm to 0.6 nm with a meager Van der Walls force in between the layers. With the use of the exfoliation process and augmented by sonification, a one-atom layer of graphite oxide soluble in water is obtained. The obtained graphite oxide will be reduced with the help of thermochemical, electromagnetic and electrochemical laser techniques.

- *Chemical Vapor Deposition*

The method of chemical vapor deposition is an innovative approach with a huge potential of patterning the monolayer of graphene into multiple layers of graphene nanoribbons. In the CVD technique, graphene wafers are patterned and fabricated onto a single polycrystalline metallic surface using pyrolysis of hydrocarbon persecutors such as methane at a high temperature. Depending on the solubility of carbon of the substrate, the graphene layers are obtained accordingly. Metals like nickel offer higher solubility of carbon, thereby triggering the ability of carbon atoms to dissolve at a greater temperature and precipitate on the metal surface, generating single or multiple layers of film of graphene. The resultant obtained films on nickel-based substrate are of non-uniform thickness ranging between 1 and 10 layers with an approximate diameter value in micrometers. The concentration of hydrocarbons and the speed of cooling determine the thickness and crystal order.

- *Chemical Synthesis*

The method of chemical synthesis is basically an organic synthesis technique. Here, the graphene can be obtained by intercalating polycyclic aromatic hydrocarbon, which is a 2D segment of graphene. This strategy is preferred due to its good flexibility and compatibility with a variety of organic synthesis techniques. Stable colloidal graphene quantum dots were recently generated using a chemical route based on benzene consisting of 132, 168 and 170 conjugated carbon atoms. The shortcoming of this technique is the reduced solubility of the graphene dots obtained by this technique.

Besides the previously mentioned techniques, a huge quantity of GNRs can be obtained using graphite nanotomy. Nano graphite is manufactured by applying a sharp-edged knife, and then these nano blocks are exfoliated, generating GNRs. With the use of H_2SO_4 (sulphuric acid) and $KMNO_4$ (potassium permanganate), multilayers of nanotubes of carbon are isolated. Another method for fabricating graphene nanoribbons is the plasma etching of nanotubes, partially implanted into the polymer film. Vacuum or laser annealing on a SiC substrate and ion implantation methods are also used to manufacture GNRs.

Properties of Graphene

The obtained graphene is then patterned into monolayers and the suspended sheets of graphene are then studied by transmission electron microscopy. The intrinsic properties of graphene like electronic, optical, structural and thermal properties were analyzed.

Structural Properties of Graphene

Graphene is a tightly packed material arranged in the form of an array of periodic C–C atoms in the dedicated sp^2-hybridized orbital. Graphene has a π bond and triple σ bonds. Last, a p_z electron forms the pie bond and this electron is responsible for the freely moving charge carrier. TEM has been used to examine the atomic structure of a single-layer graphene on the bulk of the metallic grid. The graphene sheets are suspended inbetween these metal grids.

Electronic Properties of Graphene

The graphene's electronic spectrum reveals that the charge electrons propagating the hexagonal lattice tends to lose their effective mass, thus creating a quasi-particle illustrated by the 2D equivalent of the Dirac equation, unlike the Schrodinger equation. It is observed that for low energies, the electrons' behavior is fairly massless. The holes and electrons are termed Dirac fermions and their respective six corners are called Dirac points. The experimental data of transport of graphene at ambient temperature points out that graphene has a high electron mobility (15,000 $cm^2\ V^{-1}s^{-1}$).

Thermal Properties of Graphene

The research studies show that graphene acts as a thermal conductor with a greater value of thermal conductivity (>5,000 W $m^{-1}k^{-1}$) at an ambient temperature in comparison to diamond and carbon nanotubes. The presence of propagating elastic waves into a graphene lattice determines its extensive behavior. The thermal properties are being extensively studied, yet they are in their naïve stage.

Graphene Nanoribbon Field-Effect Transistor

The field-effective transistors fabricated using GNRs as channel material are termed GNRFETs. GNRFET-based repeaters show superiority over Si-based FETs due to replacement of GNRs as channel material on the transistors (Chen et al., 2008). Although graphene exhibits zero bandgap but if patterned into nanoscale ribbons, a bandgap opens due to lateral quantum confinement (Han et al., 2007). These nanoribbons overcome short channel effects encountered in Si MOSFET. One of the important touchstones for logic circuit designing is high on/off current ratio, reduced sub-threshold swing and low-conductivity off-state for reduced power dissipation. Studies conducted (Wang et al., 2008) prove that large-diameter CNTs exhibit high current density, but low on/off current ratio; while small-diameter CNTs exhibit high on/off ratio and low current density. Like small-diameter CNTs, the GNR devices with widths of 2–3 nm ribbons exhibit high on/off current ratio.

As aforementioned, graphene nanoribbon field-effective transistor can be modeled by replacing the conventional Si channel by a GNR channel and providing metallic contacts at the source and drain regions. GNRFETs display many remarkable properties similar to CNTFETs, such as compatibility with a high-*k* dielectric, better carrier mobility for ballistic nature of transport, superior thermal conductivity and huge carrier velocity for abrupt switching. The two-dimensional graphene shows zero bandgap semimetal traits; sequencing it into nanoribbons of a few nanometer-width regimes, the bandgap opens. The graphene nanoribbon edges consist of dangling bonds that can be passivated by the action of hydrogen atoms. This action helps in determining whether the GNRs are armchair or zigzag. However, variability and certain defects manifested due to the minute thickness of nanometer-wide ribbon geometries may affect the performance of GNRFETs in general. The issue of variability may erupt from the difficulty in controlling the width of GNR, line edge roughness (LER) while doing fabrication and oxide thickness.

Types of Graphene Nanoribbon Field-Effect Transistor

Based on the device structure, GNRFETs are broadly classified as GNR-MOSFETs and GNR-Schottky Barrier FETs (GNR-SBFET). The initial graphene-made transistor framework that was experimentally and theoretically reported was a GNR-Schottky Barrier type FET (GNR-SBFET); eventually numerous other topologies have been studied that have unleashed the variety of GNRFET structure types such as Doped Channel GNRFET, Lightly Doped Drain and Source (LDDS GNRFET), Single Gate (SG-GNRFET), Double Gate (DG-GNRFET), Asymmetric Gate (AG-GNRFET), Electrically Activated Source Extension (ESE-GNRFET), Dual Material Gate (DMG-GNRFET), Two Different Gate Insulators (TDI-GNRFET) and Extra Peak Electric Field (EPF-GNRFET). Among the various GNRFET topologies discussed in the preceding section, GNR-MOSFETs are preferred for repeater circuit modeling because, compared to other types of GNRFET, the GNR-MOSFETs has a higher on/off current ratio, more robust to process variation and exhibit thermionic conduction.

Metal Oxide Semiconductor Type (MOS-GNRFET)

In MOS-type GNRFETs, the reservoirs are made to be doped with acceptors or donor impurities. By using heavily doped GNRs, ohmic contacts are obtained for the source region and drain region formation. The operational working is similar to a conventional CMOS structure.

To study the simulation behavior of the device Chen et al. (2015) proposed a parameterized, HSPICE compatible model for GNRFET. Figure 3.10(a-b) shows the structure of a MOS-GNRFET and its equivalent circuit model, respectively.

As illustrated from the GNRFET structure shown in Figure 3.10(a), the channel region is composed of a semiconducting type of graphene nanoribbons. The undoped semiconducting ribbons are arranged beneath the gate region while the highly doped ribbons are placed under the drain and source. The device can be turned on and off by applying potential on the gate terminal. The threshold voltage, V_t, required to switch on the GNRFET is mathematically given as (Illinois University GNRFET model website)

$$V_t = \frac{BG}{3e} = \frac{2\,|\delta|\,\Delta E}{3e} \tag{3.22}$$

$$\Delta E = \frac{\pi\ h\ v_f}{W_d} \tag{3.23}$$

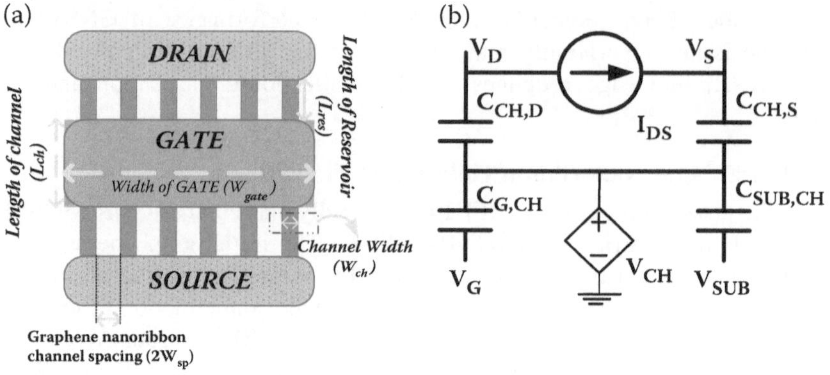

FIGURE 3.10 (a) MOS-GNRFET structure; (b) equivalent SPICE circuit model of GNRFET.

$$W_d = \frac{\sqrt{3}\,a_0(N+1)}{2} \tag{3.24}$$

where BG represents the bandgap, e denotes charge due to a unit electron, $\delta = 0.27,\ \ 0.4\ \textit{and}\ 0.066$ for $N = 3l,\ \ 3l + 1\ \textit{and}\ \ 3l + 2$, respectively. N is the number of dimer lines, $h\,(=6.5821 \times 10^{-16} eV)$ is Planck's constant and $v_f\,(=10^6 m/s)$ and W_d are thickness of GNR, or in other words width of ribbon. From Eqs. 3.22 and 3.24, we may infer that the width of ribbon, W_d, varies directly with a change in dimer lines, N, and varies inversely with bandgap, BG.

The equivalent SPICE circuit model, as shown in Figure 3.10(b), can be segregated into three parts, consisting of channel potential denoted by V_{CH}, current source I_{DS} and combination of channel capacitances denoted by variables $C_{G,\ CH}$, $C_{SUB,\ CH}$, $C_{CH,\ D}$ *and* $C_{CH,\ S}$. Moreover, from Landauer-Buttiker formalism, the value of the drain-to-source current is given as

$$I_{DS} = \frac{2e}{h} \sum\nolimits_{\propto} \int_0^{\infty} [f(E - E_{FS,C}) - f(E - E_{FD,C})]dE \tag{3.25}$$

where E is the energy level and $\propto$ is the sub-band index.

It is noted that the switching speed of the GNRFET is quite high compared to CMOS FET because of the small effective mass of carriers in graphene, leading them to respond at higher clock frequencies. Moreover, GNRFET shows better dynamic power performance by channel scaling than CMOS FETs. With an increase in the GNR width, the GNRFET shows a rise in the off-current and a reduction in the threshold voltage because of a small bandgap and a large number of available conducting sub-bands in the carrier transport. In addition to this, increasing the GNR width also increases the on-current and switching speed of GNRFET. Narrower GNRs show a high I_{ON}/I_{OFF} ratio at the expense of higher intrinsic gate-delay time as the smaller number of sub-bands are available for carrier transport, and their effective masses are higher than wider GNRs.

Thus, MOS-type GNRFETs displays better characteristics with greater on-current to off-current ratio because it does not manifest ambipolarity, better transconductance, satisfactory saturation behavior and rapid switching ability. Despite all of this, there are certain limitations with MOS-type GNRFET due to its constraint in fabrication. MOS-GNRFETs need additional doping at the source and drain regions, with a line edge roughness value (LER=0) the SS = 66.67 mV/decade and on-current to off-current ratio ($I_{ON}/I_{OFF} = 1.8 \times 10^5$) at $V_{DD} = 0.5$ V. While Si-CMOS has SS = 93.46 mV/decade and on-current to off-current ratio (I_{ON}/I_{OFF} ratio $= 3.49 \times 10^3$) at $V_D = 0.7$V. However, for the value of LER = 0.1, the SS rises to 140 mV/decade and on-current to off-current ratio (I_{ON}/I_{OFF}) drops to 9.8×10^1 at $V_{DD} = 0.5$ V.

Schottky Barrier Type (SB-GNRFET)

Out of the two most widely used topologies for GNRFETs (MOS-GNRFET and SB-GNRFET), Schottky Barrier Type is one of the most fabricated (Wessely et al., 2012).

Both types of GNRFETs can be optimized in terms of the count of nanoribbons and oxide thickness. Juxtapose to CNTFETs, both of them show superior large-area scalability along with ease due to planar fabrication possibility; but all of this is obtained at the expense of scattering effects due to edge roughness. For GNRFETs, the electronic properties and the bandgap is governed by the width of graphene nanoribbons and also the orientation corresponding to the graphene crystal, which is similar to the chirality in the case of carbon nanotube-based FETs. Pedagogical data reveals that GNR has an energy gap with an inverse proportionality to their ribbon width. In the case of Schottky Barrier Type GNRFETs topology, as shown in Figure 3.11, graphene acts as the base for intrinsic channel material while drain and source are metallic contact that leads to the formation of a Schottky Barrier at the graphene-metal interfaces.

The current conduction is primarily governed by Schottky Barrier tunneling. The trump card of using a device structure made up of Schottky Barrier Type GNRFETs is that there is no need to carry out additional doping in the metallic contacts or channel regions. This lessens the technical complexity encountered during fabrication

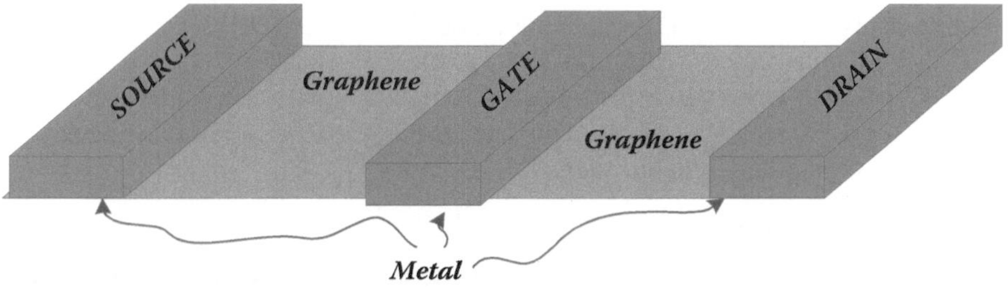

FIGURE 3.11 SB-GNRFET structure with metallic drains and source.

of devices and also eradicating problems relevant to doping variations. But, unfortunately, the ambipolar behavior of SB-GNRFETs causes performance limitations by offering a lower value for on-current to off-current (I_{ON}/I_{OFF}) ratio as compared to its counterpart MOS-type GNRFETs. In the case of a metal oxide semiconductor type (MOS-GNRFET), the drain and source reservoirs are built up of graphene. The reservoirs in MOS-GNRFETs are heavily doped with donors or acceptors forming *N*-type or *P*-type GNRFETs with *n-i-n* or *p-n-p* doping profile. The channel region is also made up of graphene in MOS-GNRFETs. Determined by the doping for reservoirs, the current in MOS-GNRFETs is dominated by electron or hole conduction and is mostly based on thermionic conduction. Thus, MOS-GNRFETs show a monotonic I–V curve because of doping in the reservoirs. The minority carriers are absorbed by dopants in a way that a large current is not created. Selection of type of dopants will determine the creation of *N*-type/*P*-type MOS-GNRFETs. In contrast to this, Schottky Barrier GNRFETs display ambipolar I-V characteristics. On presuming a mid-gap Schottky Barrier, SB-GNRFETs are obtained with a minimum current at $V_{GS}= 1/2V_{DS}$. The reason behind it is the symmetrical Schottky Barrier profiles of both holes and electrons imposed by this voltage application. In order to improve the drive strength of the device, it is advised to connect multiple GNRs in parallel fashion in GNRFET. Undoped GNR segments are placed underneath the gate, while heavily doped GNR segments are placed in between the gate and the source/drain terminals. Doped regions are the reservoirs while intrinsic regions are channels. Research studies point out that although MOS-GNRFETs offer a better I_{ON}/I_{OFF} ratio and are more resilient to the impact of process variation, SB-GNRFETs offer the benefit of not introducing high-contact resistance on the vias and a larger I_{ON}. As SB-GNRFETs display an ambipolar I-V curve, they are neither *N*-type nor *P*-type by nature. With the help of additional work function engineering, the current-voltage curve for SB-GNRFETs can be made to shift in a manner such that they act as either *N*-type nor *P*-type. In context to the I–V curve, shifting very limited practical techniques are available (Yoon et al., 2008). These techniques might be unable to shift the curve by an arbitrary amount, resulting in unbalanced *P*-type and *N*-type characteristics. To date, the finest achieved shifting in the case of SB-type transistors was ≈ 0.25V for *P*-type and –1.0V for *N*-type, using Pd and Al as gates. If the case of *N*-type and *P*-type FETs become exceedingly unbalanced, then the consequence of it is that it causes the circuit to become less robust or sometimes even nonfunctional. Henceforth, to undergo a fair comparative analysis of SB-GNRFET circuits with respect to other technologies, a conjecture related to perfect balanced shifting for both *P*-type and *N*-type transistors is considered. On the one hand, MOS-type GNRFETs can be switched on or off in a better fashion due to its greater I_{ON}/I_{OFF} ratio compared to SB-GNRFET circuits. However, with proper shifting of the I-V curve, SB-GNRFET circuits offer a higher I_{ON}. Due to such facts, MOS-GNRFETs are considered appropriate for digital circuit applications.

Doped Channel GNRFET

In an attempt to improve the performance, the doping of *P*-type and *N*-type is done in different parts of the channel. The propagation of free charge carriers increases in the channel because of the doping profile. The doping of the *N*-type improves the on-current because, as compared to holes, the electrons are lighter and have faster movement. The consequence of doping is the enhancement of impurity carrier concentration and gate voltage, thereby reducing the barrier height. The impact of barrier height reduction is a hike in on-current I_{ON} value as compared to value obtained for lightly doped devices, whereas the value for off-current I_{OFF} remains the same. The lower value of I_{ON} indicates that there is a formation of barrier near the contact interface, which can be subdued by carrying out the process of *N*-type doping. An *N*-type doped channel displays better sub-threshold swing and superior switching characteristics compared to a *P*-type doped channel. For the situation of the step doping process when there is high doping of drain region and light doping of source region, the drain becomes a good collector (as consequence of high doping there is reduction in barriers) while the source region becomes a bad emitter (due to large barriers as a consequence of lightly doped profile). The current in this case goes up with an elevated slope and good saturation condition. The doping concentration is 1: 10^3 for the 0 to 3 nm region; for *N*-type doping, the value for off-current $I_{OFF} = 3.1 \times 10^{-6}$ A and value for on-current is $I_{ON} = 3.7 \times 10^{-6}$ A. In the same way, if the doping concentration is 3: 10^3 with a 6–9 nm region, for *N*-type doping, the on-current value $I_{ON} = 9.9 \times 10^{-6}$ A and the value for off-current $I_{OFF} = 1.3 \times 10^{-6}$ A.

Lightly Doped Drain and Source (LDDS GNRFET)

One of the drawbacks of graphene nanoribbon FETs is the issue of leakage currents due to band-to-band tunneling that results in ambipolar conduction. In order to resolve this issue, it is suggested to use either a double-gate graphene nanoribbon FET (DG GNRFET) or a Lightly Doped Drain and Source graphene nanoribbon FET (LDDS GNRFET). Use of a LDDS GNRFET will result in a superior sub-threshold swing. In the case of LDDS GNRFET, a lightly doped region is fabricated inbetween the intrinsic channel and the highly doped drain and source regions. This consequence of this technique is that it reduces the probability of tunneling between the source or drain regions and the channel, thereby widening the barriers between regions. Due to reduced current spectrum in the structure, the quantum transmission also is suppressed. Thus, it is a strong contender compared to conventional GNRFET because of its better sub-threshold swing (SS) and reduced ambipolar characteristics. The sub-threshold swing value for LDDS-type GNRFETs is SS = 131.58 mV/decade, whereas without LDDS the sub-threshold swing value is SS = 254.66 mV/decade. Due to a smaller switching time, LDDS-type GNRFETs show a better I_{ON}/I_{OFF} ratio and reduce SS, reduced power delay product value and smaller delay in comparison to normal GNRFET.

Single Gate (SG-GNRFET)

Single-gate-type GNRFET, as the name identifies, contains one gate that maneuvers the current flow in the channel. The single gate uses the Schottky Barrier GNRFET. Due to the presence of a single controlling gate to direct the current flow, SG-type GNRFETs in comparison to double-gate GNRFETs, are more prone to short channel effects such as drain-induced barrier lowering. Furthermore, its saturation behavior is also inferior to double-gate GNRFETs. It exhibits a higher cutoff frequency due to reduced control over channel. It also suffers from the ambipolar conduction issues. With a reduction in the length of the channel, the on-current I_{ON} value rises while the off-current value is lower in the case of SG-GNRFETs and has a direct relationship with barrier potentials. There is a rise in the sub-threshold swing value with the reduction in length of gate. In SG-GNRFET, the value of gate capacitance and transconductance is less than DG-GNRFETs. For all such reasons, the single-gate SG-GNRFETs are preferred for low-frequency applications.

Double Gate (DG-GNRFET)

The structure of double-gate graphene nanoribbon FETs comprises of a gate surrounding the channel. Compared to single-gate SG-GNRFETs the double-gate graphene nanoribbon FETs are more effective because the gate surrounds the channel and thereby provides efficient control of the channel over the DIBL effect. Out of the two gates, the function of one gate is to control the injection of carriers and the function of the other gate is to make the band edge profile nearly flat. The value of transconductance is greater for DG-GNRFETs than SG-GNRFETs due to superior control of the gate. As a result of an extremely short length of channel, the device offers smaller delay along with a high cut-off frequency. Thus, for this reason the double-gate graphene nanoribbon FETs are suitable for the high-frequency applications and good ballistic transport. The sub-threshold swing is a function of threshold voltage, V_{th}, and also depends on the ratio of depletion capacitance and oxide capacitance, which in turn depends on the device dimensions. It is usually preferred to obtain a minimum sub-threshold swing for devices operating at a low bias voltage to eradicate the short channel effects. If the gate oxide thickness is increased, then the value of oxide capacitance decreases and the value of sub-threshold swing increases. Ideal DG-GNRFETs show 29.19% less total power consumption compared to traditional Si-CMOS. The energy delay product (EDP) of ideal DG-GNRFETs show 1% less value than Si-CMOS; despite the fact that in some cases due to line edge roughness (LER) issue, there is a decrease in the I_{ON}/I_{OFF} ratio that might determine the EDP.

Asymmetric Gate (AG-GNRFET)

The double-gate structure was initially proposed in order to curb the issue present in the case of the Schottky Barrier (SB-GNRFET) related to parasitic tunneling of current that might interfere with the ballistic transport of the device. But the complexity associated with fabricating an extra gate in the case of DG-GNRFET makes it obsolete for high-frequency digital applications. An asymmetric gate emerges as a better alternative to it, with no fabrication intricacy. Asymmetric gates are modeled to cover certain parts of the channel that are closer to the source. On reversing the voltage polarity, the asymmetric gate graphene nanoribbon FETs can be made to operate as a *P*-type device. The thickness of the Schottky Barrier in the case of an asymmetric gate graphene nanoribbon FET, at the drain end is placidly affected by the applied potential at the gate terminal. At the drain side, the asymmetric structure of the gate influences the tunneling of the hole through the Schottky Barrier. This tunneling is proportional to V_D. Reduced V_D exhibits a flatter band diagram at the drain end contact, thereby suppressing the impact of tunneling. In comparison to conventional Si-CMOS devices, the asymmetric gate graphene nanoribbon FETs (AG-GNRFETs) offer a sub-threshold swing value (SS) of 86.96 mV/decade and the ratio of on-current to off-current (I_{ON}/I_{OFF}) equals 3.04×10^2 at $V_{DD} = 0.5$ V while the normal Si-CMOS devices offer a sub-threshold swing value (SS) of 93.46 mV/decade and the ratio of on-current to off-current (I_{ON}/I_{OFF}) equals 3.49×10^3 at $V_{DD} = 0.7$ V. But for the situation when line edge roughness (LER) equals 0.1, then SS for an asymmetric structure rises to 197.24 mV/decade and the on-current to off-current (I_{ON}/I_{OFF}) ratio drops to 9.98 at $V_{DD} = 0.5$ V.

Electrically Activated Source Extension (ESE-GNRFET)

This electrically activated source extension graphene nanoribbon FET (ESE-GNRFET) shown in Figure 3.12 has a gate terminal comprising two parts (Naderi, 2015). One part, called the main gate, controls the channel voltage and the other part, called the side gate, serves as an electrically activated source extension. The side gate creates a step in the potential profile of the device and acts as a source region by the application of a suitable voltage. The structure of ESE-GNRFET has a source side of ESE and a doped section. So as to let the potential profile adjoining the junction of the source to remain flat, the length of the doped section is meticulously chosen. This length is usually kept larger in comparison to the depletion layer. For the short channel devices, if there is an increase in the source and drain voltages, the height of the potential barrier is suppressed and accordingly brings an alteration in the threshold voltage. At the source extension, for the increased biasing of the gate, a screening effect takes place, resulting in comparatively less change in the potential profile. The further consequence of this is reduction in the drain-induced barrier lowering (DIBL) effect. To obtain low threshold voltages, it is desired to have a smaller sub-threshold swing (SS). The sub-threshold swing is a function of the minority carrier density on the surface potential, which in turn is directly dependent on the gate voltage. Superior to the control of the gate voltage, the more efficient will be the performance. The sub-threshold swing (SS) is a function of selected bias points; thus, the higher the value of drain-source bias, the greater will be the sub-threshold swing. For the case of electrically activated source extension graphene nanoribbon FETs (ESE-GNRFETs), the sub-threshold swing

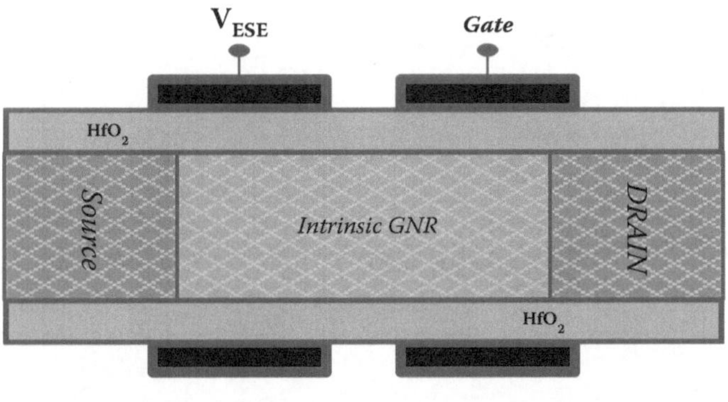

FIGURE 3.12 Electrically activated source extension (ESE) graphene nanoribbon FET.

has a reduced value in the case of a lower source extension length. There is an explicit demonstration of short channel effects when a smaller bandgap is shaped inside the channel part. The smaller the bandgap, then the wider the region for the effect of BTBT. Ambipolar current is also suppressed if a longer length of electrically activated source extension graphene nanoribbon FET is used. The impact of leakage current is reduced to a larger extent at a negative gate-source bias; due to the rise in ESE regions, it further extends the distance between the conduction band and the valence band in the tunneling region and diminishes the probability of tunneling of the device.

Dual Material Gate (DMG-GNRFET)

The dual material gate (DMG) has the gate bifurcated into left and right sides and each side has different work functions, as illustrated in Figure 3.13. There is a reduction in band-to-band tunneling due to the potential profile obtained because of a difference in the work function. This also causes a lowering of leakage currents. The advantage of using a dual material gate with different functions is the creation of an extra barrier to restrict the tunneling effect of a charge carrier moving from the valence band towards the conduction band of the drain. The step created in the potential profile provides an obstruction to the electric field from reaching the channel and thus mitigates the short channel effects like drain-induced barrier lowering. In general, if there is a rise in V_D, then the barrier is lowered, but in the case of dual material gate (DMG) graphene nanoribbon FETs, the potential step created prevents the lowering of the barrier, which in turn leads to a reduction of the drain-induced barrier lowering. At the joint of both gates, there is an extra peak in the electric field that causes an increment in the device lifetime. Due to a smaller field near the drain, there is a reduction in hot carrier effects. As compared to traditional GNRFETs, the dual material gate (DMG) graphene nanoribbon FET has a higher saturation current, lower leakage current and better ratio of on-current to off-current (I_{ON}/I_{OFF}).

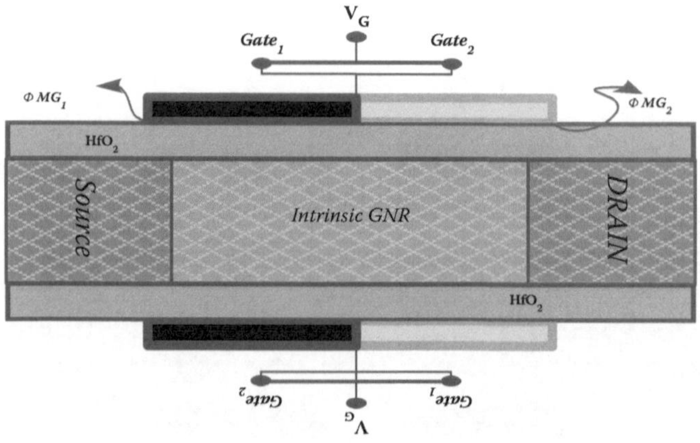

FIGURE 3.13 Dual material gate (DMG) graphene nanoribbon field-effect transistor.

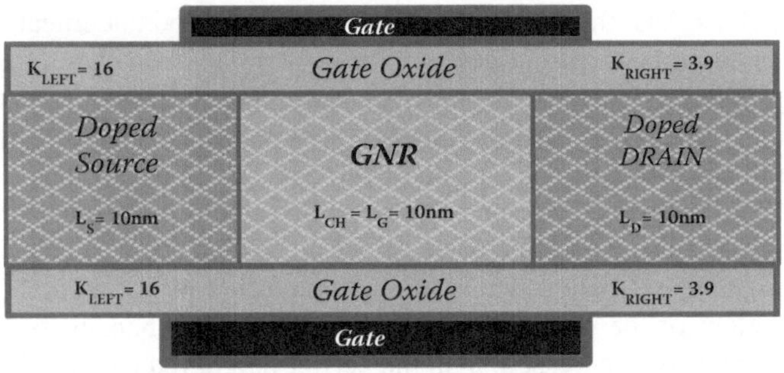

FIGURE 3.14 Two different gate insulator (TDI) graphene nanoribbon field-effect transistor.

Two Different Gate Insulator (TDI-GNRFET)

The device gets its name from its structure shown in Figure 3.14. The TDI-type GNRFETs consist of gate insulators fabricated using both high as well as low dielectrics located on the left and right side, respectively, as dielectric constant materials (Eshkalak et al., 2015). This is a type of dual gate device modeled with an aim to focus on oxide scaling. Employing a dielectric constant of low value lowers the overall parasitic capacitance. In an attempt to suppress the capacitance, the gate dielectric from the drain side (right side) is judicially selected to have a lower value in order to enhance the on-current and suppress the leakage current components. This device exhibits lower off-current value and a better ratio of I_{ON}/I_{OFF} as compared to conventional GNRFETs. Due to thermal emission current stemming as a consequence of better control on the gate potential channel near the source, there will be an increase in transconductance. As a consequence of a significant impact of the gate potential on the channel region, the drain and source capacitance are lesser at high-*k* GNRFET and larger at low-*k* GNRFET.

Extra Peak Electric Field (EPF-GNRFET)

An extra peak electric field GNRFET (EPF-GNRFET) structure is shown in Figure 3.15 and consists of dual gates, namely virtual gate (VG) and original gate (OG) (Akbari Eshkalak and Anvarifard, 2017). On applying a fixed voltage (V_{CS}) in

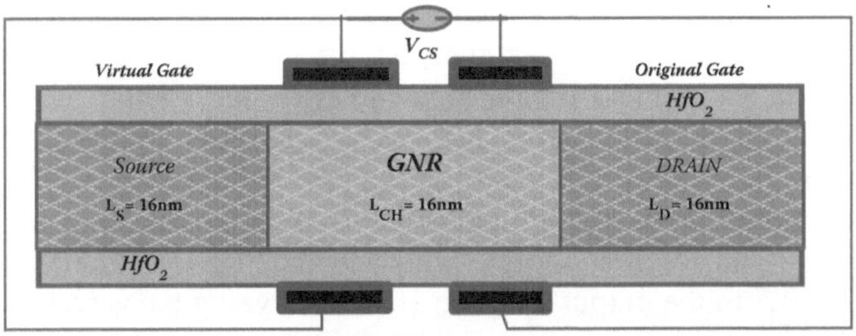

FIGURE 3.15 Extra peak electric field GNRFET (EPF-GNRFET).

between the OG and VG, then an inversion channel is created underneath the virtual gate, which assists in providing a better control over the channel. As the geometry illustrated in Figure 3.15 helps in providing modulation of surface potential distribution, hence, the result is in branching for an extra field peak electric field in the channel. Due to the mentioned effect, the device is named extra peak electric field GNRFET (EPF-GNRFET). The surface potential created beneath the original gate is quite uniform and it notably decreases under the virtual gate. This leads to a rise of an electric source extension, which can be controlled by the virtual gate. It determines the height of the potential barrier controlling the channel. As the dimension of the virtual gate is increased, then the barrier height reduces and increases the thermionic emission. In the case of extra peak electric field GNRFET (EPF-GNRFET), both the virtual gate as well as the original gate possess the same work function and are controlled by same voltage source (V_{CS}), which is applied in-between them. The transport of a charge carrier in this device is mainly governed by three factors. Firstly, because of the thermionic current, this is due to the carrier transport above the channel barrier. Secondly, it is due to the tunneling current from the source to the drain through a potential barrier. Thirdly, it is due to band-to-band tunneling inbetween conduction band and valence band (Table 3.3).

3.5 PERFORMANCE ANALYSIS OF STATE-OF-THE-ART DSM BUFFERS

In an attempt to tune the performance of the buffer driven on-chip interconnects a suitable state-of-the-art buffer driver models are paramount in order to achieve reduced path delay without undermining the power specs. The main focus of this section is to compare the performance of the state-of-the-art buffer inserted as a smart repeater inbetween interconnects to optimize the delay and power specs besides providing thermal stability. The behavioral performance analysis of graphene-based logic circuits is undertaken and it is juxtaposed with a traditional one in terms of performance metrics like delay, power and temperature variations.

3.5.1 Delay Analysis

Amidst the last few decades, the focus of researchers has moved more towards optimizing delay incurred due to a passive interconnect structure than due to active devices. With a reduction in feature size in the DSM node, the delay of active devices has been mitigated but at the cost of a rise in passive interconnect delay.

Before proceeding with the delay analysis, we will briefly present some delay time definitions in the context of a typical inverter input/output waveform, illustrated in Figure 3.16. To ease out the analysis step, a pulse input with zero rise and fall times is considered.

Propagation delay time, τ_{pd}: Propagation delay is usually the most relevant metric of interest, and is often simply called delay. It is the time from the input signal crossing 50% to the output crossing 50%. Analyzing the waveform shown in Figure 3.16, the propagation delay can be stated as the average time required for the input signal to propagate through the buffer, and thus mathematically expressed as

TABLE 3.3
Types of GNRFET structures

S.No	Structure Topologies	I_{ON}/I_{OFF} ratio	SS (mV/decade)	L_{CH} (Channel length)	V_{DS} (V)	Permittivity (ε_r)	t_{OX}(nm)	GNR	Remarks
1	MOS type GNRFET	1.8e+05	66.67	16 nm	0.5	3.9	0.95	AGNR (N = 12)	n-type doping
2	Schottky Barrier type GNRFET	3.37e+1	133.51	16 nm	0.5	3.9	1	AGNR (N = 12)	(source = 0.01.drain = 0.001 dopant/atom
3	Doped Channel GNRFET			9 nm	0.5	3.9	1.5	AGNR (N = 9)	
4	Lightly doped drain and source GNRFET		131.58	10 nm	0.5	3.9	1.5	AGNR (N = 12)	
5	Single-gate GNRFET	1.54e+04		15 nm	0.5	3.9	2	AGNR (N = 12)	
6	Double-gate GNRFET	6.7e+04		15 nm	0.5	3.9	2	AGNR (N = 12)	
7	Asymmetric gate GNRFET	3.04e+02	86.96	16 nm	0.5	3.9	2	AGNR (N = 12)	
8	Electrically activated source extension GNRFET	>2e+05		16 nmL_{ESE}=10 nmL_G=20 nm	0.5	3.9	2	AGNR (N = 13)	
9	Dual material gate type GNRFET	5823		10 nm	0.6	3.9	2	AGNR (N = 13)	
10	Two different gate insulators (TDI-GNRFET)		128.61At K_{LEFT} =16K_{LEFT}=3.9	10 nm			1.5	AGNR (N = 12)	
11	Extra peak electric field (EPF-GNRFET)		17511284	L_{CH} =12L_{CH} =16L_{CH} =20	0.6	3.9	1.5	AGNR (N = 12)	

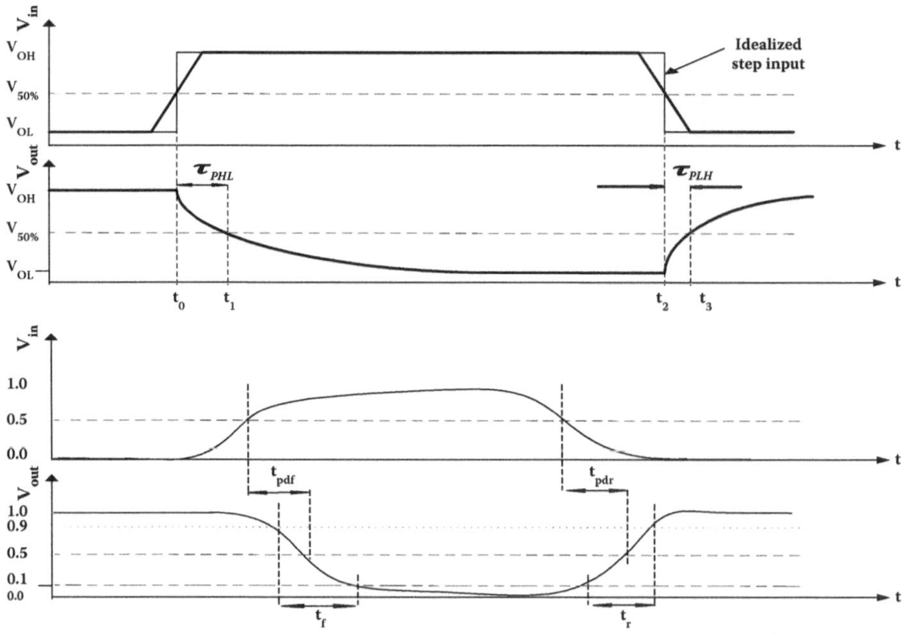

FIGURE 3.16 Inverter switching characteristics and interconnect delay effects.

$$\tau_{pd} = \frac{\tau_{PHL} + \tau_{PLH}}{2}$$

Here, τ_{PHL} is the time delay inbetween the $V_{50\%}$ transition of the rising input voltage and the $V_{50\%}$ transition of the falling output voltage, while τ_{PLH} is the time delay inbetween the $V_{50\%}$ transition of the falling input voltage and the $V_{50\%}$ transition of the rising output voltage.

Rise time, τ_{rise}: It is the time for an output waveform to rise from 10% to 90% of its steady-state value.

Fall time, τ_{fall}: It is the time for an output waveform to drop from 90% to 10% of its steady-state value.

The buffer gate that charges or discharges a node is called the driver buffer and the wire or repeater gate being driven is called the load.

Buffer Delay Estimation

The basic technique adopted for estimation of τ_{PLH} and τ_{PHL} is emanate on the calculation of the average current, I_{avg}, required to charge up and charge down the load capacitance, C_{load}. The analytical expressions are recapitulated below

$$I_{avg,HL} = \frac{1}{2}[i_C(V_{in} = V_{OH}, \quad V_{out} = V_{OH}) + i_C(V_{in} = V_{OH}, \quad V_{out} = V_{50\%})] \quad (3.26)$$

$$I_{avg,LH} = \frac{1}{2}[i_C(V_{in} = V_{OL},\quad V_{out} = V_{50\%}) + i_C(V_{in} = V_{OL},\quad V_{out} = V_{OL})] \quad (3.27)$$

$$\tau_{PHL} = \frac{C_{load}.\Delta V_{HL}}{I_{avg,HL}} = \frac{C_{load}.(V_{OH} - V_{50\%})}{I_{avg,HL}} \quad (3.28)$$

$$\tau_{PLH} = \frac{C_{load}.\Delta V_{LH}}{I_{avg,LH}} = \frac{C_{load}.(V_{50\%} - V_{OL})}{I_{avg,LH}} \quad (3.29)$$

Although the average-current technique is quite simple with less analytical calculations, it is also capable of providing an approximate first-order estimation of the charge-up and charge-down delay times. Yet it is not adopted because it ignores the change in capacitance current between the starting and end points of the transition and thus does not estimate the accurate value of delay times.

A more accurate estimation of propagation delay can be obtained by solving the below-mentioned differential equation relating to the output voltage with time; the solution obtained is termed transient response. The delay can be estimated as the time when the output reaches $\frac{V_{DD}}{2}$.

$$I = C_{\text{load}}\frac{dV_{\text{out}}}{dt} \quad (3.30)$$

The propagation delay obtained upon solving the differential equation while assuming FETs in the buffer as constant current source, is expressed as

$$\tau_{pd} = \frac{C_L(V_{DD}/2)}{I_{DSAT}} \quad (3.31)$$

The expression obtained determines a simple step response delay model; however, usually in many instances the voltage applied at the input is not a step function but rather a transient voltage pulse with a certain rise time. Due to this rise time, a current less than the full I_{DSAT} flows and consequently the propagation delay is affected by this.

Interconnect Wire Delay Estimation

The customary technique adopted for estimating the switching speed of the buffer is laid on the postulate that the nature of loads connected at the output is capacitive and lumped. As per this technique, the output load can be classified as (i) internal parasitic capacitances of FETs, (ii) interconnect wire line capacitances and (iii) input capacitances of the fan-out gates. Among these three components, the interconnect load imposes stern difficulties in modeling nano circuits. In major cases, the estimation of loading effect imposed by interconnect structures is quite trivial. As already discussed in Chapter 2, interconnect is a three-dimensional structure with distributed RLC parasitic. Hence, the delay can be calculated using any one of the two succeeding models.

RC Delay Model

It is the simplest model based on the postulate that interconnects are a network with a single lumped resistance and a single lumped capacitance. The analytical expressions are recapitulated below to compute the propagation delay using the RC delay model, assuming that the input is a rising step pulse at time t = 0.

$$V_{out}(t) = V_{DD}\left(1 - e^{-\frac{t}{RC}}\right) \tag{3.32}$$

$$V_{50\%} = V_{DD}\left(1 - e^{-\frac{\tau_{PLH}}{RC}}\right) \tag{3.33}$$

$$\tau_{PLH} \approx 0.69\,RC \tag{3.34}$$

Elmore Delay Model

The Elmore delay model (Elmore, 1948) calculates the propagation delay from a source switching to one of the leaf nodes changing as the sum over each node i of the capacitance C_i on the node, multiplied by the effective resistance R_{is} on the shared path from the source to the node and the leaf.

$$\tau_{PD} = \sum_i R_{is}\ C_i \tag{3.35}$$

3.5.2 Power Analysis

Power dissipation is yet another important issue to be dealt with carefully in interconnects modeling. With the advent of nanometer era, because of higher clock frequencies and exponential rise of on-chip integration levels, power consumption has extensively increased. Studies carried out by Gonzalez and Horowitz (1996) prove that out of the total on-chip power consumption, one-third of the microprocessor power is consumed by the clock, another third by memories and the remaining third by logic and interconnect wire structures. As a matter of fact, Magen et al. (2004) prove that dynamic power due to wire capacitance can exceed up to 50% of the total dynamic power. Additionally, as mentioned before, buffers are inserted inbetween interconnect lines as repeaters and results in dynamic, leakage and short-circuit power (Chen and Friedman, 2006). The deleterious effect of high power dissipation is an abrupt rise in packaging cost due to heating problems that consequently lead to a reduction in battery life of portable and wearable devices.

Henceforth, the modeling of portable handheld devices definitely needs contemplation for the power dissipation in order to ensure proper circuit operation and reliability. Chiefly there exist four major sources of power consumption (such as switching power, short-circuit power, leakage power and static power) in the case of any nano regime integrated circuit. The average power can be analytically expressed as the sum of all four sources of power dissipation and given as

$$P_{avg} = P_{switchng} + P_{shortckt} + P_{leakage} + P_{static} \tag{3.36}$$

$$P_{avg} = \propto_{0\to 1} C_L V V_{DD} f_{clk} + I_{sc} V_{DD} + I_{leakage} V_{DD} + I_{static} V_{DD}$$

$P_{switchng}$ denotes the switching component of power, C_L is load capacitance, f_{clk} is clock frequency and $\alpha_{0\text{->}1}$ denotes the node transition activity factor and is determined by the average number of times a node makes power-consuming transitions in a clock period. Generally the voltage swing V is taken equal to the supply voltage V_{DD}. P_{shortckt} is because of the direct-path short circuit current (I_{sc}) flowing when both N-type FETs and P-type FETs are active concurrently, thereby creating a direct path from supply voltage to ground. P_{leakage} is because of leakage current (I_{leakage}) resulting from reverse bias diode currents and sub-threshold condition governed by fabrication technology considerations. Lastly, the static current (I_{static}) flows in circuits like bias circuitry, pseudo-logic families have a steady source of current inbetween the power supplies. The succeeding section will analyze all four components in detail.

Firstly, the switching component of power comes into effect when energy is drawn from the power supply in order to charge parasitic capacitors (comprising of gate, diffusion and interconnect wire capacitance). For the circuit shown in Figure 3.17, the dynamic or switching portion of power dissipation arises if the V_{out} is either charged by the power supply or can be discharged through the ground. For the transition at the output of circuit from 0 to V_{DD}, the dynamic component of energy drawn from the power supply is expressed as $C_{\text{load}}.V_{DD}^2$. Here, the physical load capacitance at the output node is C_{load} and the potential supply applied is V_{DD}. This component of energy is predominantly independent of the function being performed and the slope or rise times for the input signals applied at the gate of transistors and solely depends on the output load capacitance and the power supply voltage. In order to calculate the average energy dissipation per switching event for the inverter shown in Figure 3.17, let us assume an input transition that causes the transition at the output of circuit from 0 to V_{DD}. For this condition, the instantaneous power and energy drawn is analytically determined as

$$P(t) = \frac{dE}{dt} = i_{supply}\ V_{DD} \tag{3.37}$$

where i_{supply} is the instantaneous current being drawn from the supply voltage V_{DD} and is expressed as

$$i_{supply} = C_L \frac{dV_{out}}{dt} \tag{3.38}$$

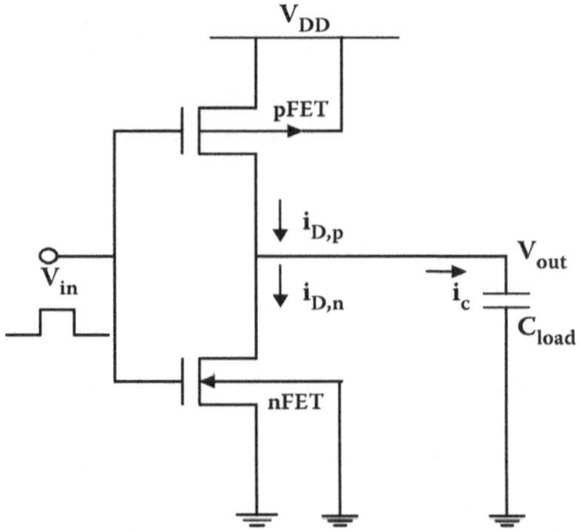

FIGURE 3.17 Buffer Inverter with lumped output load capacitance.

Henceforth, the energy drawn from the power supply for a transition from 0 to V_{DD} at the output node is expressed as

$$E_{0\to 1} = \int_0^T P(t)dt = V_{DD}\int_0^T i_{supply}(t)dt = V_{DD}\int_0^{V_{DD}} C_L\, dV_{out} = C_L V_{DD}^2 \quad (3.39)$$

Thus, it is evident from Eq. 3.39 that irrespective of the transition waveform at the output of inverter gate, the in energy drawn from the supply power for the switching transition of 0 ->V_{DD} at the output is $C_L V_{DD}^2$. The stored energy for this transition in the output capacitive load is denoted by E_{cap} and can be expressed as

$$E_{cap} = \int_0^T P_{cap}(t)dt = \int_0^T V_{out} i_{cap}(t)dt = \int_0^{V_{DD}} C_L V_{out} dV_{out} = \frac{1}{2} C_L V_{DD}^2 \quad (3.40)$$

Thence, it can be inferred from Eq. 3.40 that out of the total energy from the power supply, only 50% of the energy is stored in the output capacitive load and the remaining half of the energy is consumed in a *P*-type pull up sub-network. In the case of high to low transition (V_{DD} ->0) at the output node, there is no drawing of charge from the power supply and the stored energy in the capacitive load, which is $\frac{1}{2} C_L V_{DD}^2$, will be completely consumed in the pull down *N*-type sub-network ($E_{1\to 0} = 0$).

As mentioned in the former sections, for each transition from low to high (0 ->V_{DD}), the energy drawn at the output of the inverter gate is $C_L V_{DD}^2$. If for every clock cycle, a single transition from 0 -> V_{DD} is done at a rate of f_{clock}, then the consumed power will be $C_L V_{DD}^2 f_{clock}$. But for the condition when the rate of node transition is lower than f_{clock}, this is usually not the case; in fact, it can be greater as well.

So as to manage statistically the variation in the transition rate, take into consideration N clock periods and assume n (N) be the count of 0 ->V_{DD} transitions at the output for the time interval [0, N] in Figure 3.17. Hence, the overall energy drawn from the supply for the said interval is represented by E_N and is expressed as

$$E_N = C_L\ V_{DD}^2\ n(N) \tag{3.41}$$

In the case of an extended period of time, the average power dissipation relates to the average count of switching transitions and is expressed as

$$P_{avg} = \lim_{N\to\infty} \frac{E_N}{N} f_{clock} = \left(\lim_{N\to\infty} \frac{n(N)}{N}\ C_L V_{DD}^2 f_{clock} \right) \tag{3.42}$$

In (3.57) the limit term denotes the average (expected) value of the count of transitions per clock cycle or the node transition activity factor represented by $\propto_{0\to1}$

$$\propto_{0\to1} = \lim_{N\to\infty} \frac{n(N)}{N} \tag{3.43}$$

Thus, the average power can now be given as

$$P_{avg} = \propto_{0\to1}\ C_L V_{DD}^2 f_{clock} \tag{3.44}$$

For the case of multiple transistor gates like NAND or NOR logic, the internal node (V_{int}) of a P-type pull up sub-network might undergo transitions; thus, the transition activity should be computed for all the nodes of a given circuit. Henceforth, the complete power for a circuit can be computed by adding overall circuit nodes (i) and is recapitulated as

$$P_{total} = \sum_{i=1}^{number\ \ of\ \ nodes} \propto_i C_i V_{DD}^2 f_{clock} \tag{3.45}$$

On considering the probability of an individual node swinging to a voltage V_i, which can be lower than V_{DD}, then in that situation the total power is given as

$$P_{total} = (\sum_{i=1}^{number\ \ of\ \ nodes} \propto_i C_i V_i\ ) V_{DD} f_{clock} \tag{3.46}$$

We critically analyzed the dynamic or switching component of total power consumed in a given circuit. The dynamic power relates to the overall energy needed in order to charge the parasitic capacitors. This component of power does not depend on the rise and fall times at the input of logic gates. But the fact cannot be overlooked that a finite rise time and fall time of the input waveforms will result in a direct path for current to flow from V_{DD} to ground. This situation exists for a

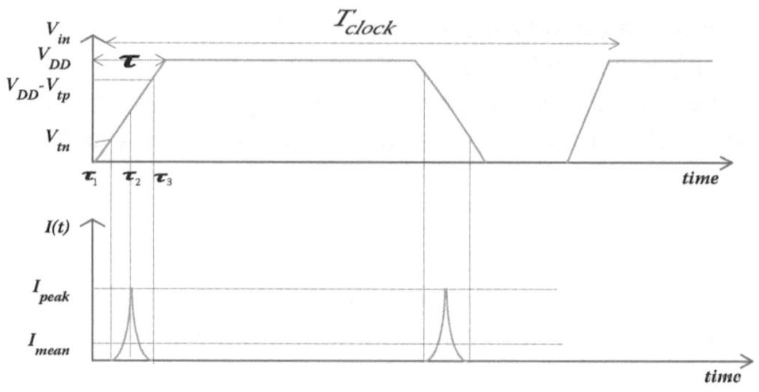

FIGURE 3.18 Short circuit current behavior.

very short span of time during switching, particularly if $V_{Tn} < V_{in} < V_{DD} - |V_{Tp}|$ exists for the input voltage, there will be a conductive path open between V_{DD} and ground as both the *N*-type and *P*-type devices are on. Such a type of path cannot exist in dynamic circuits, because precharge and evaluate transistors must never be on at the same time; otherwise, it may result in an incorrect evaluation. Henceforth, we may state that short circuit currents are an issue faced solely by static circuit designs.

Figure 3.18 illustrates the behavior of short circuit current for an inverter circuit without any output load. When there is a 0->1 transition at the input, then the *N*-type pull down sub-network starts conducting at V_{in} and becomes equal to V_{tn}; whereas a *P*-type pull up sub-network stops conducting at V_{in} equal to $V_{DD} - |V_{Tp}|$. As under the condition of zero capacitive loads, all the current that is drawn from the power supply is consumed as short circuit power. The exact analysis for the short circuit dissipation is quite trivial; hence, a simplified analysis taking into consideration that a symmetrical inverter (having $\beta_N = \beta_P$) with an effective transistor strength is presented in the following paragraph.

Assume the rise time of the input signal to be τ, and the threshold voltage (V_t) is equal for both *N*-type and *P*-type transistors (such that $V_{tN} = V_{tP}$). Let $\beta = \beta_N W_N = \beta_P W_P$. From $\tau_1 = \tau_2$, the *N*-type pull down network is in saturation as the output voltage is higher than $V_{in} - V_t$. The current flowing in the pull down sub-network is recapitulated as

$$I = \frac{\beta}{2}(V_{in} - V_t)^2 \quad for \ \ 0 < I < I_{max} \tag{3.47}$$

Because of the supposition of inverter symmetry, the current reaches its peak value at $V_{in} = \frac{V_{DD}}{2}$. The mean current is calculated analytically to find out the average short circuit current drawn from the supply:

$$I_{mean} = \frac{1}{T}\int_0^T I(t)dt = 2\frac{2}{T}\int_{\tau_1}^{\tau_2} \frac{\beta}{2}(V_{in}(t) - V_t)^2 dt \tag{3.48}$$

where $V_{in}(t) = \frac{V_{DD}}{\tau}.\ t$ and for $V_{in}(0) = 0,\ \ V_{in}(\tau_1) = V_{DD}$ *and* $V_{in}(\tau_2) = 0$; therefore, τ_1 *and* τ_2 are given by $\tau_1 = \frac{V_t}{V_{DD}}\tau$ and $\tau_2 = \frac{\tau}{2}$. On substituting these values in Eq. 3.48, we get

$$I_{mean} = \frac{2\beta}{T_{clock}}\int_{\tau/2}^{V_t\tau/V_{DD}} \left(\frac{V_{DD}t}{\tau} - V_t\right)^2 d\left(\frac{V_{DD}t}{\tau} - V_t\right) = \frac{\beta}{12V_{DD}}(V_{DD} - 2V_t)^3 \frac{\tau}{T_{clock}} \tag{3.49}$$

Therefore the short circuit power for an unloaded inverter is

$$P_{sc} = \frac{\beta}{12}(V_{DD} - V_T)^3 \frac{\tau}{T} \tag{3.50}$$

Thus, the short circuit current is directly proportional to the rise time and the effective transistor strength β. In fact, these short circuit currents are of considerable value if the rise/fall time at the input of the inverter gate is quite longer than the output rise/fall time. The reason behind this is due to the short circuit path being active for a greater span of time. Hence, by choosing an appropriate transistor size, the short circuit power can be kept to less than 10%. Moreover, operating the circuits at a supply voltage less than the sum of the *N*-type and *P*-type threshold voltages will definitely eliminate any short circuit currents.

The third factor of power is because of leakage currents. There are primarily two types of leakage currents: reverse-bias diode leakage at the transistor drains and sub-threshold leakage through the channel of an "off' device. The sub-threshold leakage occurs due to carrier diffusion between the source and the drain when the gate-source voltage, V_{gs}, has exceeded the weak inversion point, but is still below the threshold voltage, V_t, where carrier drift is dominant. In this regime, the FETs behave like a bipolar transistor, and the sub-threshold current is exponentially dependent on the gate-source voltage, V_{gs}. An important figure of merit for a low-power technology is the sub-threshold slope SS, which is the amount of voltage required to drop the sub-threshold current by one decade. The lower the sub-threshold slope the better, since the devices can be turned off as close to V_t as possible. Even though the leakage power constitutes a minute fraction of the total power dissipation in most chips, it could be noteworthy for a system application that spends much of its time in standby operation, since this power is always being dissipated even when no switching is occurring.

Lastly, the fourth component of power is due to the static currents found in circuits that do not have rail-to-rail swing feeding other circuits, such as pseudo-*N*-type logic styles, and in analog bias circuits. The static currents must be minimized as much as possible. For example, in SRAM sense amplifiers, pulsed circuits should be used to minimize static currents.

Power Dissipation Components of Repeater Inserted Interconnect Network

The following section discusses power characteristics of an interconnect network with repeater insertion. The total power consumption for buffer-driven on-chip interconnects is the sum of *dynamic power, short-circuit power* and *leakage power* in buffer.

The scaling of technology causes a growing integration of excessive functionality onto a single application-specific integrated circuit (ASIC), thereby leading to enlargement of die size despite the reduction in minimum device feature size. Furthermore, the technology scaling results in an increase count of lengthy global interconnect wires. As the delay of long buffer lines is proportional to the length of wire, hence it is suggested to segment the long wires into small portions by inserting repeater buffers inbetween. The delay of the buffer gate is a linear function of wire length. But for high-performance designs, the count of such repeater buffers might prohibitively rise and eventually might occupy a considerable fraction of active silicon and routing area. It can also cause a significant amount of total chip power consumption and rise in leakage current. As far as power consumption is concerned, a noteworthy portion of overall on-chip power dissipation is because of loading caused by interconnecting structures, particularly in high-performance designs. Many literatures report 50%–75% of power dissipation simply because of a clock circuitry distribution network. Generally, inserted power-efficient repeater buffers are optimally sized and staggered separately in order to improve the performance. In view of the fact that not all the global interconnect lines lies on the critical path (path having maximum delay), a small fraction of delay penalty can be overlooked for the case of such noncritical interconnect wires if there exists a large potential to achieve significant power savings with the use of smaller buffer repeaters and greater inter-repeater interconnect lengths.

While carrying out the power-efficient optimal repeater insertion analysis for efficient on-chip interconnects performance benchmarking, it is recommended to consider all three types of power consumption. Since in the nanometer regime VLSI interconnects, the leakage power is rising abruptly; also, the short circuit power comes out to be 25% of the total power consumption for high-speed and low-power sub-threshold VLSI circuits. Henceforth, ignoring them in performance benchmarking (while analyzing power, PDP and EDP design metrics) may result in significant errors and may risk the validity of optimized performance parameters.

The three different components of total power consumption by a repeater inserted interconnect network are expressed as follows.

Dynamic Power

The dynamic power mainly consists of the switching power. In other words a dynamic power is the amount of power requisite to charge and discharges the various devices and interconnects capacitances. To compute the power, every node of the circuit is taken into consideration. The overall capacitances of the node are the sum of the gate, diffusion and wire capacitance of the node. Hence, analytically, the total dynamic power is expressed as the summation of the power from the line capacitance and the repeaters.

$$P_{Dynamic} = P_{Dynamic-wire} + P_{Dynamic-repeater} \tag{3.51}$$

$$P_{Dynamic-wire} = C_{interconnect} * V_{dd}^2 * f \tag{3.52}$$

$$P_{Dynamic-repeater} = \alpha_{switching} * k_{opt} * h_{opt} * C_o * V_{dd}^2 * f \tag{3.53}$$

$P_{Dynamic-repeater}$ depends upon the switching factor and the number k_{opt} and size h_{opt} of each repeater. While the number of repeaters decreases, the repeater size increases. The dynamic power dissipated by a wire increases with greater wire capacitance (as the line width is increased). The dynamic power of the repeaters, however, decreases since fewer repeaters are used with wider wires. The reduction in power dissipated by the repeaters overcomes the increase in the interconnect power until the line capacitance dominates the line impedance. After exceeding a certain width, the total power increases with increasing line width. All of the terms are functions of the line width except V_{dd}, C_O and f. To optimize the power, it is good to select the minimum V_{dd} that can support the required frequency of operation.

Short Circuit Power

If the input waveform applied at the buffer repeater has a finite slew rate, then a direct current path exists from supply V_{dd} to ground at the instance the signal switches from V_{in} to $V_{dd} + V_{tp}$. The power dissipated in this manner is short circuit power (Chatzigeorgiou, Nikolaidis, and Tsoukalas, 2001). For an interconnect repeater buffer system, the narrow wires cause less dynamic power and higher short-circuit power dissipation. Thus, for thin lines the total repeater count can be large. Short-circuit power depends on input transition time, output load capacitance and FET size and is given by

$$P_{shortcircuit-wire+buffer} = \frac{1}{2} * I_{peak} t_{base} V_{dd} f \tag{3.54}$$

Here, I_{peak} is the peak current that flows from V_{dd} to ground, t_{base} is the time for which both FETs were on, V_{dd} is the supply voltage and f is the switching frequency.

Leakage Power in Buffer

For a repeater buffer in nanometer technology, the primary leakage current source consists of sub-threshold current and gate leakage current. Thus, total leakage power dissipated in repeater buffers is

$$P_{leakage} = h \ k \ V_{dd} (I_{sub0} + I_{g0}) \tag{3.55}$$

Here, I_{sub0} is the average current of FETs in minimum-sized repeaters. I_{g0} is the average gate leakage current in minimum-sized repeaters with low and high inputs. It is predicted that in the near future the leakage power will dominate dynamic and short circuit power. Although with the adoption of graphene-based FETs, the problem of leakage current is mitigated to a considerable extent.

Power Reduction Techniques

Since the dynamic power or switching power is a function of power supply voltage (V_{dd}), load capacitance (C_L), clock frequency (f_{clk}) and switching activity ($\alpha_{switching}$), suppressing these parameters can directly affect the lowering of dynamic power dissipation.

The dynamic power is a quadratic ally dependent on the V_{dd}; therefore, operating at a lower power supply voltage can considerably diminish the dynamic power consumption in the case of a repeater inserted on chip interconnect networks. However, an operating network at an excessive lower power supply voltage results in a significant rise in signal propagation delay. Therefore, the trade-off between power dissipation and propagation delay is done.

In addition to this, mitigation of load capacitance may be accomplished by either implementing smaller gates in output or considering smaller number of fan-out gates. Using a small interconnect wire length at the output can further suppress the load capacitance.

Besides this, operating at lower clock frequency and less switching activity of the design also helps in suppressing the dynamic power dissipation in the case of a repeater inserted on chip interconnect networks.

Whereas in order to mitigate the short circuit power dissipation, it is recommended to have a quick rise/fall time of the input signal, low clock frequency and low power supply voltage. The leakage power can be mitigated by either reducing the leakage current or operating at a reduced power supply voltage. Furthermore, modeling logic circuits with transistors with a multi-threshold voltage can be used to reduce leakage power at the system level and by using sleep transistors connected in series, the standby power can be reduced.

3.5.3 Electro Thermal Stability Analysis

For the faster logic devices with the rise in circuit speeds, larger power dissipation is often reported, thereby causing excessive heating of IC. With a rise in temperature, the power-delay product of MOSFET increases, whereas the power-delay product of CNTFET almost remains constant (Khursheed et al., 2019). Furthermore, the utmost leakage power of the MOSFET gates rises in a linear proportion with respect to temperature, while for graphene-based FETs the leakage power increases exponentially. It is a fact that even a slight rise in leakage power ultimately shoots the temperature, a consequence of this in the long run is a further increase of leakage power, henceforth creating a vicious cycle. Moreover, in the quest of designing ultra-low-power buffers as repeaters for on-chip interconnects the squelching of threshold voltage leads to an exponential rise in leakage power. Hence, overlooking the electrothermal effects while modeling low power-efficient buffers for nano-scale interconnects may lead to suboptimal power consumption. To overcome this issue, a sophisticated temperature-aware repeater insertion model is proposed by Banerjee et al. (2003). This model is based on a one-dimensional heat equation to estimate the chip temperature using first-order approximation and is described as

$$\theta_{ja} c_T \frac{dT}{dt} + T = (P_{repeater}(T) + P_{logic}(T))\theta_{ja} + T_a \tag{3.56}$$

The silicon substrate and package junction-ambient resistance is θ_{ja}, the thermal capacitance is denoted by c_T and the ambient temperature ($\approx 45°C$) is denoted by T_a. The chip temperature, T, is not only the function of the repeater power, $P_{repeater}$, but also depends on power dissipated in surrounding logic blocks, P_{logic}. This P_{logic} is composed of dynamic ($P_{dynamic-logic}$) and leakage power and described as

$$P_{logic} = P_{dynamic-logic} + W_{eff} I_{leakage} V_{dd} \begin{pmatrix} c_1 (T - T_0)^2 \\ + c_2 (T - T_0) + c_3 \end{pmatrix} \tag{3.57}$$

The dynamic power term is constant as it is unaffected by temperature variation and also repeater insertion methodology. The leakage power depends on temperature, effective width of gates in logic block (W_{eff}), leakage current per unit width of buffer gate at the nominal temperature ($I_{leakage}$) and technology specific constants c_1, c_2, c_3.

3.6 ARCHITECTURE AND WORKING OF REPEATER BUFFER FOR ON-CHIP INTERCONNECTS

Apart from the aforementioned material and technological advancements in device modeling of repeater buffers for on-chip interconnects; there is a lot of scope for improvement in their circuit designing (Venkatesan et al., 2000). The existing strategy is to ply basic buffer circuits as repeaters along interconnecting wires simply to boost the signal. However, these repeater buffers cost chip area and power, and their number is raising rapidly in future ICs. A trade-off among different design criteria is needed for a reliable repeater insertion methodology. Many different logic gates, implemented as buffer circuits, are briefly described in this section. These buffer repeater logic gates are designed to handle the various performance issues; nevertheless, the main problem that still exists is either high energy expenditure or extensive delay times. Although in certain circuits mentioned below, with the capability to symmetrically reduce the voltage swing on wire, relatively shorter delay times and low energy dissipation can be achieved, but with dreadful consequences of reduced noise margin.

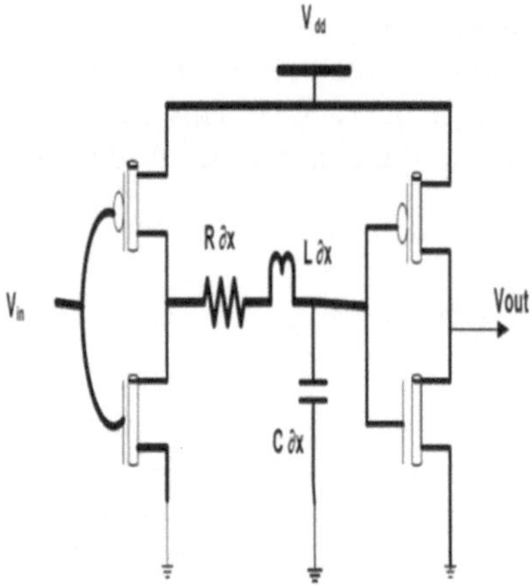

FIGURE 3.19 NOT gate as a repeater.

Basic NOT Logic Gate as Repeater

As depicted in Figure 3.19, a basic logic NOT gate inverter as a repeater uses a simple diode inverter as a buffer inbetween the lengthy interconnect lines. The inverter consists of a series combination of p-FET and n-FET. The signal propagation delay in a conventional repeater is measured as half the total sum of rising signal delay and falling signal delay. Although it reduces the overall propagation delay of on-chip interconnect paired with buffers, it increases the power dissipation (Secareanu and Friedman, 2000).

LECTOR as Repeater

As depicted in Figure 3.20, the Lector inverter technique is a well-known technique adopted for leakage power reduction (Hanchate and Ranganathan, 2004). In this approach, transistors are arranged in a fashion such that two extra (one *P*-type and

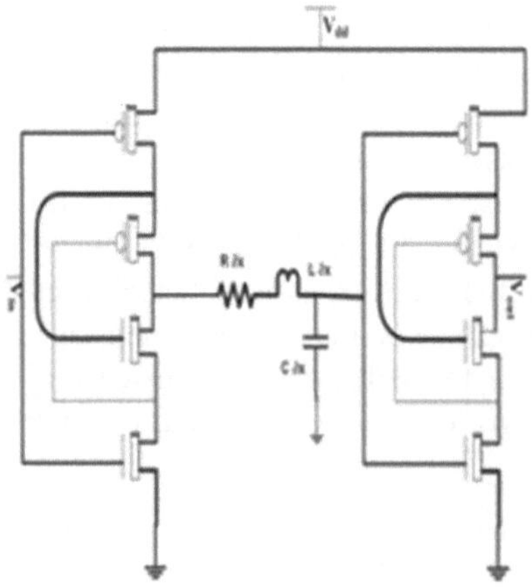

FIGURE 3.20 LECTOR as repeater.

one N-type) leakage control transistors (LCTs) are inserted in between a pull up network (PUN) and pull down network (PDN) logic circuit. The gate node of each LCT is governed by a source of another LCT. In the LECTOR architecture technique, for any input combination, one of the LCTs is always operating near the cut-off voltage. Thus, resistance of V_{dd} to ground path increases, resulting in a decrease of leakage current.

Schmitt-Trigger as Repeater

As shown in Figure 3.21, the Schmitt-trigger is a suitable alternative for replacement of conventional buffers as repeaters. It offers the choice of adjustable threshold voltage and hysteresis. Thus, keeping a lower threshold value of the Schmitt-trigger signal can rise early and the large noise margin of the Schmitt-trigger helps in reducing the noise glitches as well (Saini et al., 2009).

Current Mode Logic (CML) Buffer as Repeater

Current mode logic (CML) shown in Figure 3.22 is a variant of source coupled differential logic, which is preferred for designing high-speed circuits. Both CNT-based CML and MOS-based CML comprise of three sections: a differential PDN, a pull up load circuit and constant bias current source. The load transistors are made to operate in a linear region, whereas the current source NFET transistor and PDN transistors are made to operate in a saturation region, thereby resulting in minimum source voltage. CML operates on the principle that PDN switches the constant current between two branches and the load converts current to output voltage swings. During low to high, the PUN charges the output to V_{dd} and high to low NFET charges the output to $V_{dd} - I_B R_D$. Here, V_{dd} is a source voltage, I_B is a bias

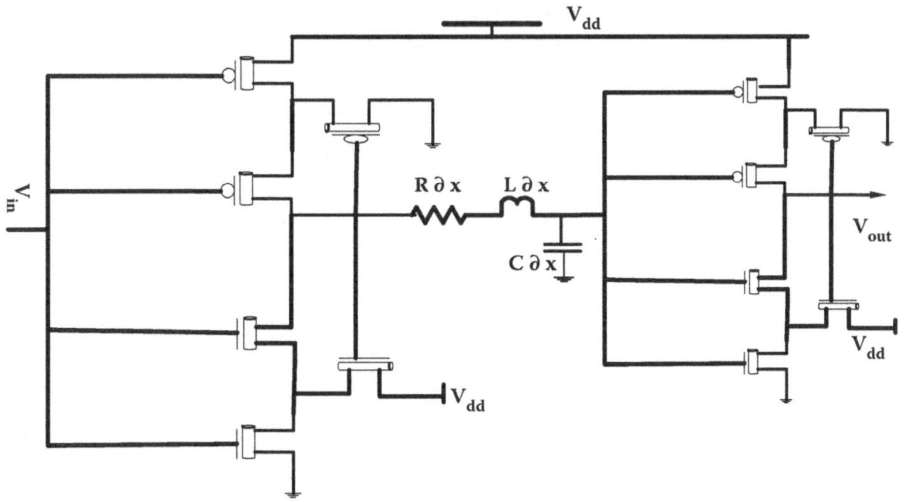

FIGURE 3.21 Schmitt-trigger as repeater.

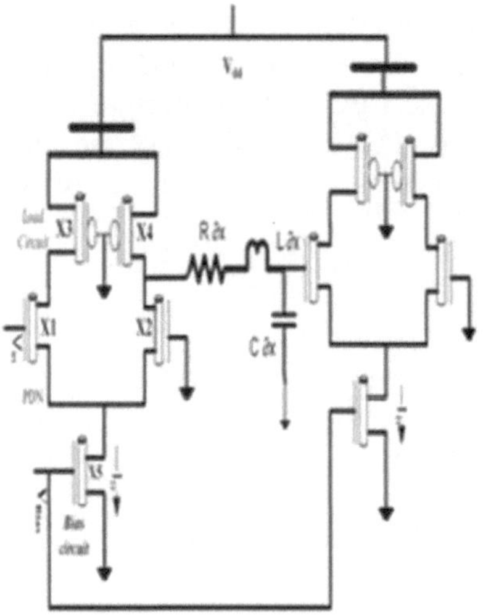

FIGURE 3.22 Current mode logic (CML) buffer as repeater.

current and R_D is a load impedance of PFET. The output logic swing is $\Delta V = I_B \times R_D$. Due to a lower voltage swing, the circuit results in lower delays. Power consumption of CML is independent of the operating frequency and is expressed as a product of V_{dd} and I_B. The static power in CML gates is high compared to other buffers due to constant power consumed in circuits during and after switching of output nodes (Thorat et al., 2016). In our test buffers, we overcome this limitation by lowering the power dissipation using different power gating strategies.

3.7 CROSS-TECHNOLOGY PERFORMANCE BENCHMARKING OF REPEATER BUFFER FOR ON-CHIP NANO INTERCONNECTS

In this segment, we qualitatively assessed and investigated the performance of CNTFET as well as GNRFET repeater buffers against the conventional Si-CMOS 45 nm and 90 nm process technology-based different logic gates, which are implemented as repeater buffers inbetween on-chip nano interconnects. The benchmarking is done for the performance metrics like cut-off frequency ($f_T = \frac{1}{2\pi}\frac{g_m}{C_G + C_s + C_d + C_{sub} + C_{int} + C_{ext} + C_w}$), drain-induced barrier lowering ($DIBL = \partial V_T/\partial V_{DS}$), sub-threshold swing ($SS = \partial V_{GS}/\partial log_{10} I_{DS}$), on-current to off-current ratio (I_{ON}/I_{OFF}), power delay product ($PDP = P_{average} \times t_p$) and energy delay product ($EDP = PDP \times t_p$) for the device having a threshold (V_T), drain to source voltage (V_{DS}), gate to source voltage (V_{GS}), drain current (I_{DS}), average power ($P_{average}$) and propagation delay (t_p). The circuit-level simulations are conducted on an HSPICE simulator platform using the Stanford compact SPICE model for CNFET (Deng and Wong, 2007), SPICE-compatible compact models of MOS- and SB-GNRFETs from (Chen et al., 2013a, 2013b) and BSIM4 predictive technology model for CMOS.

For carrying out the device performance benchmarking of carbon nanotubes and graphene nanoribbon FETs in opposition to silicon-based scaled CMOS FETs, it is necessary to ensure a fair assessment. In order to achieve the correct result, both carbon-based devices as well as silicon devices are assessed for analogous current drive strength. The CNTFET/GNRFET dimensions of a channel are compared in lieu with Si CMOS FETs to obtain values for parameters like SS and DIBL from the drain to gate characteristics plot of the device. It is to be noted that output drain current plays a critical role in determining the transistor switching speed for logic gates. For the same current output vicinity, it is observed for CNTFETs/GNRFETs, a notable reduction in the values of sub-threshold swing and also drain-induced barrier lowering as compared to short and long channel nanoscale Si CMOS FETs, whereas the on-current to off-current ratio (I_{ON}/I_{OFF}) of the CMOS FETs channel is quite better than its carbon complement. Table 3.4 enlists key parameters to benchmark the performance of 50 nanometer CNTFET as well as 20 nanometer GNRFETs against 45 nanometer and 200 nanometer gate lengths Si CMOS FETs, respectively, for a 45 nanometer technology process node.

TABLE 3.4
Performance benchmarking parameters for CNTFET, GNRFET and MOSFET

Parameter	CNTFET Benchmarking		GNRFET Benchmarking	
	CNTFET	CMOS FET	GNRFET	CMOS FET
Channel Length	50 nm	45 nm	20 nm	200 nm
Channel Width	–	125 nm	–	120 nm
Contact Width	100 nm	–	100 nm	–
Effective Channel Area	5×10^{-15} m^2	5.63×10^{-15} m^2	2×10^{-15} m^2	2.4×10^{-14} m^2
CNT diameter./GNR Width	1.5437 nm	–	2.2 nm	–
Chiral Vector [n,m]	[20,0]	–	[19,0]	–
Maximum drain Current	46.56 μ A	50.20 μ A	19.92 μ A	25.40 μ A
Transconductance	68.1 μ S	148 μ S	27.98 μ S	63.8 μ S
Carrier Density	30.16 μA/nm	0.40 μA/nm	9.05 μA/nm	0.21 μA/nm
Load Capacitance	46.54fF	50.13 f F	19.9 fF	25.1 fF
Gate Capacitance	14.85 a F	65.85 aF	5.55 aF	269.60 aF
Drain Capacitance	0.59 aF	19.0 aF	0.54 aF	18.60 aF
Source Capacitance	1.43 aF	78.7 aF	0.22 aF	267.00 aF
Substrate Capacitance	1.60 aF	6.52 aF	0.71 aF	28.50 aF
Total Terminal Capacitance	18.47 aF	209.02 aF	7.01 aF	619.70 aF
Wire Capacitance (5 μm)	783.7 aF	783.7 aF	783.7 aF	783.7 aF
Intrinsic Capacitance	21.29 aF	37.40 aF	12.29 aF	36.10 aF
Extrinsic Capacitance	44.07 aF	384.0 aF	16.48 aF	1190 aF
Total Capacitance	867.5 aF	1414.12 aF	819.48 aF	2629.5 aF
Cutoff Frequency	12.49 GHz	16.65 GHz	5.43 GHz	3.86 GHz
DIBL	40.85 mV/V	83.89 mV/V	40.91 mV/V	115.2 mV/V
Sub-threshold swing	72.3 mV/dec	113.67 mV/dec	70.20 mV/dec	111.7 mV/dec
I_{ON}/I_{OFF} ratio	2.99×10^4	9.54×10^6	3.08×10^4	4.08×10^6

From Table 3.4, the following inferences can be drawn:

- The FETs fabricated using nanotubes of carbon displays cutoff frequency lower by 25% than the Si CMOS FETs of 45 nm gate length; while the cutoff frequency in the case of graphene nanoribbons FETs is significantly higher than the Si CMOS FETs by 40%. Moreover, as the cutoff frequency is inversely proportional to interconnect length, it is found that if the length of the interconnect wire is kept larger than 10 μm, then irrespective of the process technology node the frequency remains unchanged. Henceforth, it is recommended to prefer dividing the long wires into small segments by inserting buffer repeaters inbetween them. By doing the decimation of long interconnect wires, it is possible to exploit the high-frequency capability of the CNTFETs as well as GNRFETs (Nougaret et al., 2009; Lin et al., 2010). From the obtained results, state-of-the-art CNTFETs and GNRFETs demonstrate operating frequencies up to 80 GHz and 100 GHz, respectively.
- The drain current obtained using CNTFETs is in the same range as obtained by Si CMOS FETs of 45 nm gate length.
- Both CNTFETs as well as GNRFETs show promising values of DIBL and SS. The drain-induced barrier lowering effects and sub-threshold swing are remarkably suppressed in the case of CNTFETs/GNRFETs, whereas the Si CMOS FETs exhibit a moderate value of SS mainly because of short channel effects.
- Despite the fact that CNTFETs and GNRFETs demonstrate analogues on-current value, still it sustains the on-current to off-current ratio (I_{ON}/I_{OFF}) two orders of magnitude lower than Si CMOS FETs.

Table 3.5 recapitulates the physical feature dimensions (channel width W, channel length L and area A) of Si CMOS FETs, CNTFETs and GNRFETs on the basis of 45 nm and 90 nm technology process.

The tabulated data of Table 3.5 clearly points out that by implementing FETs with nanotubes of carbon, a total of 11% mitigation in area is accomplished in the case of an *N*-type CNTFET and for *P*-type CNTFET it is 36% of t_{node} = 45 nm juxtapose to Si CMOS FETs. However, the quintessential nanoribbon FETs demonstrate the highly efficient area consumption by resulting in 90% mitigation of area for both the *P*-type as well as *N*-type GNRFETs. In addition to these, quasi-ballistic GNRFETs have a maximum drain current, I_{ds}, value roughly around 20 μA. In order to achieve similar current value using Si CMOS FETs (i.e. in the range of 20–25 μA), the length of channel should be increased from the minimum gate feature size.

Propagation delay is an important parameter of thought that measures the performance of an on-chip nano device. As discussed in Section 3.5.1, it is the time taken by any signal to propagate in an interconnecting wire with interpolated buffers as a repeater inbetween. The signal latency of interconnect depends on the length of the interconnect. The interconnect wire is considered a distributed RLC transmission line. The propagation delay is a quadratic function of length and is given by the Elmore delay and RC delay. For the simulation purpose, CNT- and GNR-based logic buffer circuits are operated using standard HSPICE models compatible with 45 nm and 90 nm Si CMOS FETs. To achieve a fair assessment, Si

TABLE 3.5
Physical feature dimensions of Si CMOS FETs, CNTFETs and GNRFETs for a 45 nm and 90 nm process technology

Type	t_{node} = 45 nm								
	MOSFET			CNTFET			Percentage Change		
	W	L	A	W	L	A	Δ W	Δ L	Δ A
N-FET	125 nm	45 nm	5.63 fm^2	100 nm	50 nm	5 fm^2	−0.20	+0.11	−0.11
P-FET	175 nm	45 nm	7.88 fm^2	100 nm	50 nm	5 fm^2	−0.42	+0.11	−0.36
	MOSFET			GNRFET			Percentage Change		
	W	L	A	W	L	A	Δ W	Δ L	Δ A
N-FET	120 nm	200 nm	24 fm^2	100 nm	20 nm	2 fm^2	−0.17	−0.90	−0.92
P-FET	140 nm	200 nm	28 fm^2	100 nm	20 nm	2 fm^2	−0.29	−0.90	−0.93
				t_{node} = 90 nm					
Type	MOSFET			CNTFET			Percentage Change		
	W	L	A	W	L	A	Δ W	Δ L	Δ A
N-FET	120 nm	200 nm	24 fm^2	220nm	50 nm	11 fm^2	+0.83	−0.75	−0.54
P-FET	270nm	200 nm	54 fm^2	220 nm	50 nm	11 fm^2	−0.19	−0.75	−0.80
	MOSFET			GNRFET			Percentage Change		
	W	L	A	W	L	A	W	L	A
N-FET	120 nm	500nm	60 fm^2	220 nm	20 nm	44 fm^2	+0.83	−0.96	−0.93
P-FET	250nm	500nm	125 fm^2	220 nm	20 nm	44 fm^2	−0.12	−0.96	−0.96

CMOS FETs are modeled for high-current and low-current designs to compare with CNTFET compact models and GNRFETs, respectively. To extract the Si CMOS FETs parameters, a schematic of a CMOS-based logic buffer repeater gate is created in Cadence and then the layout of gate is drawn followed by a design rule check (DRC) step. Next, LVS check using Assura is done. After verifying the layout design, parasitics are extracted using Assura RCX. The obtained values of extracted parasitics are then fed into a circuit simulation software environment to achieve complete device simulation. Thus, the logic operation at this stage is simulated using a tool to scrutinize the propagation delay of the repeater buffer logic gate. Besides this, the effective delay for buffer gates can be analytically computed with the help of Eqs. 3.3–3.10 and Eqs. 3.28–3.35. To estimate and extrapolate the electrical impedance parameter values of an on-chip interconnect, a detailed mathematical formulation is carried out in Section 2.5 of Chapter 2.

Table 3.6 enlists interconnect Cu wire capacitance values of 1 μm and 5 μm lengths for 45 nm and 90 nm technology process node. Table 3.7 enlists diverse thickness (100 μm and 500 μm) for the value of substrate insulator capacitance for CNTFET and

TABLE 3.6
Interconnect Cu wire capacitance values of 1 μm and 5 μm length for 45 nm and 90 nm technology process node

Technology process node	Interconnect Cu wire capacitance	
	1 μm	5 μm
45 nm	157.8 aF	782.6 aF
90 nm	183.8 aF	924.7 aF

TABLE 3.7
Substrate capacitance for 100 μm and 500 μm of CNTFET and GNRFET

Type of Device	Substrate capacitance	
	100 μm	500 μm
CNTFET (L = 50 nm)	1.9510 aF	1.5130 aF
GNRFET (L = 20 nm)	0.7058 aF	0.6934 aF

TABLE 3.8
Resistance for CNT (20,0) and GNR (19,0)

Process parameters	CNT	GNR
Chiral Vector	(20,0)	(19,0)
Length	50 nm	20 nm
R_Q	6.453 kΩ	12.906 kΩ
R_{nc}	3.231 kΩ	1.365 kΩ
$R_{contact}$	9.681 kΩ	14.271 kΩ
$R_{channel}$	3.225 kΩ	17.208 kΩ
R_{on}	12.906 kΩ	31.479 kΩ
G_{on}	G_0	0.41 G_0
m.f.p.	100 nm	15 nm

GNRFET. Table 3.8 enlists CNT and GNR for the values of different simulation process parameters including the contact channel as well as quantum resistances.

To benchmark the performance in terms of propagation delay metrics, the transit delay for Si CMOS FETs, CNTFETs and GNRFETs is calculated for an interconnect length of 45 nm technology node design guidelines and also for 90 nm technology node design guidelines. Tables 3.9, 3.10 and 3.11 enlist the calculated delay values for Si CMOS FETs, CNTFETs and GNRFETs, respectively. The

TABLE 3.9
Si CMOS FETs delay computation

CMOSFETs buffer	For comparison with CNTFET (L = 45 nm)					For comparison with GNRFET (L = 200 nm)				
	P-type width (nm)	*N*-type width (nm)	τ_{PLH}(ps)	τ_{PHL}(ps)	Delay (τ_{pd}) (ps)	P-type width (nm)	N-type width (nm)	τ_{PLH}(ps)	τ_{PHL}(ps)	Delay (τ_{pd}) (ps)
					t_{node} = 45 nm					
Basic NOT	375	125	6.672	6.689	6.6805	290	120	19.05	18.91	18.98
LECTOR,	475	125	8.710	8.680	8.695	140	270	20	20.06	20.03
Schmitt- trigger	175	260	5.734	5.722	5.728	140	215	14.41	15.04	14.725
CML	175	125	3.663	3.71	3.6865	140	120	9.235	9.181	9.208
					t_{node} = 90 nm					
CMOSFETs buffer	For comparison with CNTFET (L = 200 nm)					For comparison with GNRFET (L = 500 nm)				
	P-type width (nm)	*N*-type width (nm)	τ_{PLH}(ps)	τ_{PHL}(ps)	Delay (τ_{pd}) (ps)	P-type width (nm)	N-type width (nm)	τ_{PLH}(ps)	τ_{PHL}(ps)	Delay (τ_{pd}) (ps)
Basic NOT	460	120	29.07	30.14	29.605	380	120	96.93	97.6	97.26
LECTOR	610	120	51.57	50.85	51.21	520	120	157.9	159.9	158.9
Schmitt- trigger	270	165	23.69	24.64	24.165	250	150	85.91	87.33	86.62
CML	270	120	10.42	10.49	10.455	250	120	34.81	34.46	34.635

TABLE 3.10
CNTFETs delay computation

	t_{node} = 45 nm	
CNTFETs buffer	**Delay without interconnects Delay(τ_{pd}) (ps)**	**Delay with 5 μm interconnect Delay(τ_{pd}) (ps)**
Basic NOT	0.47	12.98
LECTOR	0.61	16.87
Schmitt-trigger	0.39	12.97
CML	0.14	9.277
	t_{node}= 90 nm	
CNTFETs buffer	**Delay without interconnects Delay(τ_{pd}) (ps)**	**Delay with 5 μm interconnect Delay(τ_{pd}) (ps)**
Basic NOT	0.63	15.18
LECTOR	0.81	19.84
Schmitt-trigger	0.52	15.17
CML	0.19	11.07

TABLE 3.11
GNRFETs delay computation

	t_{node}= 45 nm	
GNRFETs buffer	**Delay without interconnects Delay(τ_{pd}) (ps)**	**Delay with 5 μm interconnect Delay(τ_{pd}) (ps)**
Basic NOT	0.48	30.48
LECTOR	0.63	39.41
Schmitt-trigger	0.40	30.19
CML	0.14	22.23
	t_{node}= 90 nm	
GNRFETs buffer	**Delay without interconnects Delay(τ_{pd}) (ps)**	**Delay with 5 μm interconnect Delay(τ_{pd}) (ps)**
Basic NOT	0.86	36.13
LECTOR	1.11	45.65
Schmitt-trigger	0.71	35.99
CML	0.26	25.99

propagation delay of CNTFET and GNRFET logic buffer repeater circuits are computed with and without considering parasitic interconnect effects. Table 3.9 shows the width feature dimensions of the *P*-type as well as *N*-type FETs, low-high propagation delay (τ_{PLH}), high-low (τ_{PHL}) and also list the simulation results of average propagation delay ($\tau_{pd} = \frac{\tau_{PHL} + \tau_{PLH}}{2}$) for logic buffer circuits such as Basic NOT, LECTOR, Schmitt-trigger and current mode logic (CML) buffer.

It is viable from the simulation results of Tables 3.10 and 3.11 that the rise and fall delays of the GNRFETs are twice that of the CNTFETs. This is due to the additional current available from a CNT and due to the valley degeneracy of 2 compared to 1 for the GNR. Nevertheless, these delays can be improved when GNRs are able to provide a higher drain current with improved contact interface. The value of transit delays is mainly governed by the resistive-capacitive elements in a circuit. The overall delay values obtained are determined from the effective values of RC delay, high to low propagation delay, low to high propagation delay and rise time and fall time. For easy analysis, the CNT- and GNR-based FETs are modeled as an effective resistive element connected in series with interconnect wire of Cu material. Here, both CNTFET- as well as GNRFET-based different buffer circuits are assessed for two distinct technology process nodes cases of 45 nm and 90 nm. Initially, these buffers are modeled using carbon-based circuit design and are simulated with 45 nm contact design rules; after that, the simulations are followed using 90 nm design rules. Here, the size of the contact plays a crucial role as it is the main factor that decides the value of parasitic capacitance present inbetween the bulk and source or drain node and also the resistance of ohm contact. As the process technology node is scaled from 180 nm to 45 nm, there has been observed a remarkable reduction in value of parasitic capacitance, primarily because of scaling in area size by 94%. This mitigation in capacitance directly or indirectly affects the delay as process technology changes from 90 nm to 45 nm. The result of Tables 3.9–3.11 clearly demonstrate that the delay for the same buffer circuit is more for the 90 nm process than the 45 nm process. Furthermore, propagation delay increases if the interconnect is considered because now interconnect capacitive effect also comes into picture. The length of interconnect wire taken also shows a profound effect on signal propagation delay. Generally for the case of digital logic simulation, average wire length of 5 μm is taken. Thus, the prime limiting factor for high-speed ASICs is the interconnect structure itself. Henceforth, the overall performance improvement achieved by using CNTFETs and GNRFETs is overshadowed if the interconnect capacitance is not judicially reduced with scaling in transistor feature size. In all three tables a similar trend is seen in case of Lector buffer delay increases drastically at the cost of a power-saving strategy adopted, while CML buffers have a minimum delay because of reduced voltage swing architecture. Compared to the basic NOT gate, the conventional CML buffer delay is suppressed by 68% and the Schmitt buffer suppresses the delay by only 17%. The switching speed of CML is improved in comparison to Lector but on the penalty of power dissipation.

Apart from Delay metrics, the performance can also be benchmarked in terms of energy delay product (EDP) and power delay product (PDP) metrics. Both the PDP and EDP are better design metrics for evaluation of such circuits, because they represent a combined measure of both energy and performance. It describes the trade-off between power and performance at a given operating frequency and is also termed figure of merit for devices. It is desired to have a minimum value of the EDP and PDP. Table 3.12 demonstrates the performance benchmarking of CNTFETs against Si CMOS FETs in terms of PDP and EDP design metrics without considering the interconnect effect, whereas Table 3.13 demonstrates the performance benchmarking of GNRFETs against Si CMOS FETs for PDP and EDP design metrics without taking interconnect structure into consideration. Both Tables 3.12 and 3.13 are assessed for 45 nm and 90 nm process design rules under the influence of substrate insulator thickness variation (100 nm and 500 nm).

The simulation results presented in Tables 3.12 and 3.13 confirm that the CNTFET buffer circuits exhibits a lower PDP than the Si CMOS FETs buffer logic. This investigation does not take into account the effects of intermediate wire capacitance. The GNRFET-based buffers show the PDP value almost doubled compared to CNTFET buffer circuits, whereas its EDP value is quadrupled compared to that of the CNTFET-based buffers. Both CNTFETs as well as GNRFETs buffer devices demonstrate extremely low value of PDP and EDP compared to the Si CMOS FET buffers by at least four orders of magnitude.

In Tables 3.14 and 3.15, examination of PDP and EDP values are done, with interconnect structure taking into consideration the wire capacitance. The investigations are carried out for both CNTFETs and GNRFETs and benchmarked against Si CMOSFETs, subjected to variation in substrate insulator thickness (100 nm and 500 nm). The simulation results are assessed for 45 nm and 90 nm technology node considering Cu interconnect wire lengths of 1 μm and 5 μm to take into account wire capacitance effect.

For simulating the benchtest architecture, we use a copper interconnect of 45 nm with a 100-nm and 500-nm substrate insulator thickness. The substrate thickness has a vital role in the result obtained for EDP and PDP design metrics, particularly in lower length domain. The tabulated result makes it pretty clear that no notable variation with substrate thickness is obtained beyond 1 μm interconnecting wire length. Analytically these figures of merit (PDP and EDP) for logic buffers can be computed by multiplying the average power consumed by the specific buffer with the effective signal propagation delay, such as PDP = $P_{avg} \times T_p$ and EDP = PDP $\times T_p$.

In addition to this, it can be interpreted from the result data that power delay products of CNTFET-based repeater buffer logic gates have significantly lower values compared to the power delay product value obtained using Si CMOSFET-based repeater buffer logic gates under similar conditions. Notably for the case of 45 nm design process, the energy delay product of carbon-based FET buffer circuits is found to be two times smaller than that of Si CMOS FET-based buffer circuits. If the length of wire is taken as 5 μm, then the PDP value of carbon-modeled transistors is a decade less than conventional CMOS-modeled transistors, whereas for without interconnect case, i.e. when length of wire is 0 μm, the value of PDP considerably increases to a thousand times, as seen from the table. For both the mentioned cases, variation in PDP and EDP is examined due to substrate insulator thickness (100 nm and 500 nm). On comparing the result for 45 nm and 90 nm technology nodes, it can be stated that energy delay product increases enormously with the progress in process technology. Focusing on two cases taken with interconnect wire consideration when substrate thickness is 100 nm and 500 nm, by simulating with 1 μm and 5 μm wire segment, the energy delay product gap

TABLE 3.12
CNTFETs PDP and EDP computation without interconnects for 45 nm and 90 nm technology node with variation in substrate insulator thickness

Types of Buffers	PDP without interconnects				EDP without interconnects			
	CNTFET (t_{sub}= 100 nm)		CMOSFETs		CNTFET (t_{sub}= 100 nm)		CMOSFETs	
	t_{node}= 45 nm	t_{node} = 90 nm	t_{node} = 45 nm	t_{node} = 90 nm	t_{node} = 45 nm	t_{node} = 90 nm	t_{node} = 45 nm	t_{node} = 90 nm
Basic NOT	1.208×10^{-20}	10.18×10^{-20}	16.90×10^{-18}	224.5×10^{-18}	7.18×10^{-33}	17.67×10^{-32}	1.49×10^{-28}	7.25×10^{-27}
LECTOR	3.45×10^{-20}	22.13×10^{-20}	25.25×10^{-18}	341.1×10^{-18}	27.98×10^{-33}	36.15×10^{-32}	2.87×10^{-28}	15.3×10^{-27}
Schmitt– trigger	1.45×10^{-20}	9.59×10^{-20}	15.53×10^{-18}	132.8×10^{-18}	7.566×10^{-33}	11.26×10^{-32}	1.35×10^{-28}	3.44×10^{-27}
CML	0.24×10^{-20}	1.14×10^{-20}	7.03×10^{-18}	50.90×10^{-18}	0.46×10^{-33}	0.51×10^{-32}	0.35×10^{-28}	0.70×10^{-27}
Types of Buffers	**CNTFET (t_{sub}= 500 nm)**		**CMOSFETs**		**CNTFET (t_{sub}= 500 nm)**		**CMOSFETs**	
	t_{node} = 45 nm	t_{node} = 90 nm	t_{node} = 45 nm	t_{node} = 90 nm	t_{node} = 45 nm	t_{node} = 90 nm	t_{node} = 45 nm	t_{node} = 90 nm
Basic NOT	8.25×10^{-21}	14.31×10^{-21}	16.90×10^{-18}	224.5×10^{-18}	3.236×10^{-33}	$7.44x10^{-33}$	1.49×10^{-28}	7.25×10^{-27}
LECTOR	19.70×0^{-21}	34.17×0^{-21}	25.25×0^{-18}	341.1×10^{-18}	12.04×10^{-33}	$27.54x10^{-33}$	2.87×10^{-28}	15.3×10^{-27}
Schmitt– trigger	7.814×10^{-21}	13.49×10^{-21}	15.53×10^{-18}	132.8×10^{-18}	3.673×10^{-33}	$8.463x10^{-33}$	1.35×10^{-28}	3.44×10^{-27}
CML	1.33×10^{-21}	2.34×10^{-21}	7.03×10^{-18}	50.90×10^{-18}	0.19×10^{-33}	$0.45x10^{-33}$	0.35×10^{-28}	0.70×10^{-27}

TABLE 3.13

GNRFETs PDP and EDP computation without interconnects for 45 nm and 90 nm technology node with variation in substrate insulator thickness

Types of Buffers	PDP without interconnects				EDP without interconnects			
	GNRFET (t_{sub} = 100 nm)		CMOSFETs		GNRFET (t_{sub} = 100 nm)		CMOSFETs	
	t_{node} = 45 nm	t_{node} = 90 nm	t_{node} = 45 nm	t_{node} = 90 nm	t_{node} = 45 nm	t_{node} = 90 nm	t_{node} = 45 nm	t_{node} = 90 nm
Basic NOT	1.079×10^{-20}	1.080×10^{-19}	62.06×10^{-18}	1188×10^{-18}	9.507×10^{-33}	3.051×10^{-31}	1.30×10^{-27}	62.06×10^{-18}
LECTOR	2.74×10^{-20}	2.64×10^{-19}	98.55×10^{-18}	1779×10^{-18}	30.86×10^{-33}	9.340×10^{-31}	2.87×10^{-27}	98.55×10^{-18}
Schmitt- trigger	1.130×10^{-20}	1.089×10^{-19}	59.61×10^{-18}	843.1×10^{-18}	8.143×10^{-33}	2.489×10^{-31}	1.27×10^{-27}	59.61×10^{-18}
CML	0.16×10^{-20}	0.17×10^{-19}	20.03×10^{-18}	298.8×10^{-18}	0.42×10^{-33}	0.14×10^{-31}	0.23×10^{-27}	20.03×10^{-18}
Types of Buffers	**GNRFET (t_{sub} = 500 nm)**		**CMOSFETs**		**GNRFET (t_{sub} = 500 nm)**		**CMOSFETs**	
	t_{node} = 45 nm	t_{node} = 90 nm	t_{node} = 45 nm	t_{node} = 90 nm	t_{node} = 45 nm	t_{node} = 90 nm	t_{node} = 45 nm	t_{node} = 90 nm
Basic NOT	3.413×10^{-21}	1.041×10^{-20}	62.06×10^{-18}	1188×10^{-18}	1.65×10^{-33}	8.99×10^{-33}	1.30×10^{-27}	62.06×10^{-18}
LECTOR	8.596×10^{-21}	2.64×10^{-20}	98.55×10^{-18}	1779×10^{-18}	5.38×10^{-33}	29.25×10^{-33}	2.87×10^{-27}	98.55×10^{-18}
Schmitt- trigger	3.586×10^{-21}	1.09×10^{-20}	59.61×10^{-18}	843.1×10^{-18}	1.43×10^{-33}	7.74×10^{-33}	1.27×10^{-27}	59.61×10^{-18}
CML	0.58×10^{-21}	0.174×10^{-20}	20.03×10^{-18}	298.8×10^{-18}	0.083×10^{-33}	0.458×10^{-33}	0.23×10^{-27}	20.03×10^{-18}

TABLE 3.14
CNTFETs PDP and EDP computation with Cu interconnects of 1 μm and 5 μm for 45 nm and 90 nm technology node with variation in substrate insulator thickness

Types of Buffers	PDP with 1 μm interconnects				EDP with 1 μm interconnects			
	CNTFET (t_{sub} = 100 nm)		CMOSFETs		CNTFET (t_{sub} = 100 nm)		CMOSFETs	
	t_{node} = 45 nm	t_{node} = 90 nm	t_{node} = 45 nm	t_{node} = 90 nm	t_{node} = 45 nm	t_{node} = 90 nm	t_{node} = 45 nm	t_{node} = 90 nm
Basic NOT	0.31×10^{-18}	0.70×10^{-18}	16.90×10^{-18}	224.5×10^{-18}	9.98×10^{-31}	3.25×10^{-30}	1.49×10^{-28}	7.25×10^{-27}
LECTOR	0.46×10^{-18}	1.24×10^{-18}	28.65×10^{-18}	681.7×10^{-18}	17.7×10^{-31}	6.68×10^{-30}	3.34×10^{-28}	39.4×10^{-27}
Schmitt- trigger	0.32×10^{-18}	0.73×10^{-18}	15.53×10^{-18}	132.8×10^{-18}	9.45×10^{-31}	3.05 × 10–30	1.35×10^{-28}	3.44×10^{-27}
CML	0.17×10^{-18}	0.29×10^{-18}	7.03×10^{-18}	50.90×10^{-18}	3.19×10^{-31}	0.69×10^{-30}	0.35×10^{-28}	0.70×10^{-27}
Types of Buffers	**CNTFET (t_{sub} = 500 nm)**		**CMOSFETs**		**CNTFET (t_{sub} = 500 nm)**		**CMOSFETs**	
	t_{node} = 45 nm	t_{node} = 90 nm	t_{node} = 45 nm	t_{node} = 90 nm	t_{node} = 45 nm	t_{node} = 90 nm	t_{node} = 45 nm	t_{node} = 90 nm
Basic NOT	0.39×10^{-18}	0.59×10^{-18}	25.25×10^{-18}	341.1×10^{-18}	14.3×10^{-31}	2.63×10^{-30}	2.87×10^{-28}	15.3×10^{-27}
LECTOR	0.43×10^{-18}	0.63×10^{-18}	28.65×10^{-18}	681.7×10^{-18}	17.0×10^{-31}	2.99×10^{-30}	3.34×10^{-28}	39.4×10^{-27}
Schmitt- trigger	0.28×10^{-18}	0.41×10^{-18}	15.53×10^{-18}	132.8×10^{-18}	8.05×10^{-31}	1.42×10^{-30}	1.35×10^{-28}	3.44×10^{-27}
CML	0.16×10^{-18}	0.23×10^{-18}	7.034×10^{-18}	50.90×10^{-18}	2.90×10^{-31}	0.51×10^{-30}	0.35×10^{-28}	0.70×10^{-27}
Types of Buffers	**PDP with 5 μm interconnects**				**EDP with 5 μm interconnects**			
	CNTFET (t_{sub} = 100 nm)		CMOSFETs		CNTFET (t_{sub} = 100 nm)		CMOSFETs	
	t_{node} = 45 nm	t_{node} = 90 nm	t_{node} = 45 nm	t_{node} = 90 nm	t_{node} = 45 nm	t_{node} = 90 nm	t_{node} = 45 nm	t_{node} = 90 nm
Basic NOT	5.44×10^{-18}	8.54×10^{-18}	15.53×10^{-18}	132.8×10^{-18}	7.06×10^{-29}	13.4×10^{-29}	1.35×10^{-28}	3.44×10^{-27}
LECTOR	7.43×10^{-18}	12.61×10^{-18}	25.25×10^{-18}	341.1×10^{-18}	12.7×10^{-29}	26.7×10^{-29}	2.87×10^{-28}	15.3×10^{-27}
Schmitt- trigger	5.41×10^{-18}	8.45×10^{-18}	16.90×10^{-18}	224.5×10^{-18}	7.11×10^{-29}	13.7×10^{-29}	1.49×10^{-28}	7.25×10^{-27}
CML	3.77×10^{-18}	5.55×10^{-18}	7.03×10^{-18}	50.9×10^{-18}	3.53×10^{-29}	6.33×10^{-29}	0.35×10^{-28}	0.70×10^{-27}
Types of Buffers	**CNTFET (t_{sub} = 500 nm)**		**CMOSFETs**		**CNTFET (t_{sub} = 500 nm)**		**CMOSFETs**	
	t_{node} = 45 nm	t_{node} = 90 nm	t_{node} = 45 nm	t_{node} = 90 nm	t_{node} = 45 nm	t_{node} = 90 nm	t_{node} = 45 nm	t_{node} = 90 nm
Basic NOT	6.95×10^{-18}	9.82×10^{-18}	28.65×10^{-18}	681.7×10^{-18}	$11.5x10^{-29}$	19.2×10^{-29}	3.34×10^{-28}	39.4×10^{-27}
LECTOR	7.15×10^{-18}	9.99×10^{-18}	25.25×10^{-18}	341.1×10^{-18}	$12.1x10^{-29}$	19.8×10^{-29}	2.87×10^{-28}	15.3×10^{-27}
Schmitt- trigger	5.28×10^{-18}	7.31×10^{-18}	16.90×10^{-18}	224.5×10^{-18}	$6.85x10^{-29}$	11.1×10^{-29}	1.49×10^{-28}	7.25×10^{-27}
CML	3.72×10^{-18}	5.25×10^{-18}	7.034×10^{-18}	50.90×10^{-18}	$3.45x10^{-29}$	5.8×10^{-29}	0.35×10^{-28}	0.70×10^{-27}

TABLE 3.15

GNRFETs PDP and EDP computation with Cu interconnects of 1 μm and 5 μm for 45 nm and 90 nm technology node with variation in substrate insulator thickness

Types of Buffers	PDP with 1 μm interconnects				EDP with 1 μm interconnects			
	GNRFET (t_{sub} = 100 nm)		CMOSFETs		GNRFET (t_{sub} = 100 nm)		CMOSFETs	
	t_{node} = 45 nm	t_{node} = 90 nm	t_{node} = 45 nm	t_{node} = 90 nm	t_{node} = 45 nm	t_{node} = 90 nm	t_{node} = 45 nm	t_{node} = 90 nm
Basic NOT	0.89×10^{-18}	2.14×10^{-18}	89.07×10^{-18}	1803×10^{-18}	7.85×10^{-30}	2.75×10^{-29}	2.37×10^{-27}	32.7×10^{-26}
LECTOR	0.92×10^{-18}	2.17×10^{-18}	98.55×10^{-18}	1779×10^{-18}	8.22×10^{-30}	2.82×10^{-29}	2.87×10^{-27}	27.1×10^{-26}
Schmitt- trigger	0.62×10^{-18}	1.32×10^{-18}	62.06×10^{-18}	1188×10^{-18}	4.31×10^{-30}	1.34×10^{-29}	1.30×10^{-27}	12.5×10^{-26}
CML	0.38×10^{-18}	0.69×10^{-18}	20.03×10^{-18}	298.8×10^{-18}	1.73×10^{-30}	0.43×10^{-29}	0.23×10^{-27}	1.25×10^{-26}
Types of Buffers	**GNRFET (t_{sub} = 500 nm)**		**CMOSFETs**		**GNRFET (t_{sub} = 500 nm)**		**CMOSFETs**	
	t_{node} = 45 nm	**t_{node} = 90 nm**	**t_{node} = 45 nm**	**t_{node} = 90 nm**	**t_{node} = 45 nm**	**t_{node} = 90 nm**	**t_{node} = 45 nm**	**t_{node} = 90 nm**
Basic NOT	0.75×10^{-18}	1.18×10^{-18}	89.07×10^{-18}	1803×10^{-18}	6.19×10^{-30}	12.1×10^{-30}	2.37×10^{-27}	32.7×10^{-26}
LECTOR	0.78×10^{-18}	1.21×10^{-18}	98.55×10^{-18}	1779×10^{-18}	6.53×10^{-30}	12.5×10^{-30}	2.87×10^{-27}	27.1×10^{-26}
Schmitt- trigger	0.55×10^{-18}	0.83×10^{-18}	62.06×10^{-18}	1188×10^{-18}	3.44×10^{-30}	6.66×10^{-30}	1.30×10^{-27}	12.5×10^{-26}
CML	0.36×10^{-18}	0.54×10^{-18}	20.03×10^{-18}	298.8×10^{-18}	1.56×10^{-30}	2.95×10^{-30}	0.23×10^{-27}	1.25×10^{-26}
Types of Buffers	**PDP with 5 μm interconnects**				**EDP with 5 μm interconnects**			
	GNRFET (t_{sub} = 100 nm)		**CMOSFETs**		**GNRFET (t_{sub} = 100 nm)**		**CMOSFETs**	
	t_{node} = 45 nm	**t_{node} = 90 nm**	**t_{node} = 45 nm**	**t_{node} = 90 nm**	**t_{node} = 45 nm**	**t_{node} = 90 nm**	**t_{node} = 45 nm**	**t_{node} = 90 nm**
Basic NOT	15.68×10^{-18}	23.68×10^{-18}	89.07×10^{-18}	1803×10^{-18}	6.15×10^{-28}	11.3×10^{-28}	2.37×10^{-27}	32.7×10^{-26}
LECTOR	15.85×10^{-18}	23.91×10^{-18}	98.55×10^{-18}	1779×10^{-18}	6.28×10^{-28}	11.5×10^{-28}	2.87×10^{-27}	27.1×10^{-26}
Schmitt- trigger	12.37×10^{-18}	18.92×10^{-18}	62.06×10^{-18}	1188×10^{-18}	3.76×10^{-28}	7.16×10^{-28}	1.30×10^{-27}	12.5×10^{-26}
CML	8.92×10^{-18}	12.88×10^{-18}	20.03×10^{-18}	298.8×10^{-18}	2.00×10^{-28}	0.34×10^{-28}	0.23×10^{-27}	1.25×10^{-26}
Types of Buffers	**GNRFET (t_{sub} = 500 nm)**		**CMOSFETs**		**GNRFET (t_{sub} = 500 nm)**		**CMOSFETs**	
	t_{node} = 45 nm	**t_{node} = 90 nm**	**t_{node} = 45 nm**	**t_{node} = 90 nm**	**t_{node} = 45 nm**	**t_{node} = 90 nm**	**t_{node} = 45 nm**	**t_{node} = 90 nm**
Basic NOT	15.09×10^{-18}	20.51×10^{-18}	89.07×10^{-18}	1803×10^{-18}	5.84×10^{-28}	9.26×10^{-28}	2.37×10^{-27}	32.7×10^{-26}
LECTOR	15.39×10^{-18}	20.77×10^{-18}	98.55×10^{-18}	1779×10^{-18}	6.07×10^{-28}	9.48×10^{-28}	2.87×10^{-27}	27.1×10^{-26}
Schmitt- trigger	12.09×10^{-18}	17.10×10^{-18}	62.06×10^{-18}	1188×10^{-18}	3.65×10^{-28}	6.18×10^{-28}	1.30×10^{-27}	12.5×10^{-26}
CML	8.80×10^{-18}	12.18×10^{-18}	20.03×10^{-18}	298.8×10^{-18}	1.96×10^{-28}	3.16×10^{-28}	0.23×10^{-27}	1.25×10^{-26}

inbetween CNTFET-based buffer circuits reduces with an increase in wire length for both conditions.

Similarly, for GNRFET-based buffer circuits, the thickness of the substrate insulator plays a crucial role especially for a smaller segment of interconnect wire length. If the length of the interconnect wire is taken as 5 μm, then a significant impact cannot be seen on the values of energy delay product and power delay product. Henceforth, it can be deduced that the delay as well as energy efficiency of a graphene nanoribbon-based FET buffer circuits can be easily beleaguered by wire capacitance if the length is too long. The tabulated result portrays that for a 45 nm process with 500 nm substrate thickness, an improvement of 28% in power delay product value and improvement of 39% in power delay product value is observed for GNRFET buffer circuits. Thus, the PDP for GNRFETs rises by twofold compared to CNTFETs, whereas for EDP it is nearly four times greater than that of CNTFET.

In the aforementioned paragraphs, the carbon-modeled devices are benchmarked with conventional Si CMOS buffer devices under the 45 nm and 90 nm foundry technology in terms of different design metrics. From the above discussion, it can finally be concluded that both CNTFET-based buffer circuits as well as GNRFET-based buffer circuits have exceedingly low values of power delay product and energy delay product in comparison to Si CMOS buffer devices, by nearly four orders of magnitude. Moreover, the PDP of GNRFET-based buffer circuits are doubled compared to CNTFET-based buffer circuits, while there the EDP is nearly quadrupled compared to that of the CNTFET. Even while taking into consideration the interconnect, the PDP for a distribution of buffer repeater logic gates with a 5 μm wire length are nearly 50% superior than that of the Si CMOSFET. It is a noteworthy point that these estimated results can be further improved by using state-of-the-art material for the production of fine interconnects made from metallic GNR and CNT, as discussed in Chapter 2. These novel materials have resistances and capacitances much lower than copper interconnects, thus improving the performance metric of CNTFETs and GNRFETs. In addition to this, it is found that a thicker substrate insulator can help to reduce the EDP and PDP by a great deal when the interconnecting wires are kept shorter than 5 μm.

It is irrespective of the fact that insertion of buffer circuits as a repeater is a trusted technique to reduce the overall propagation delay occurred due to interconnect wire capacitances. However, putting an excess number of buffers as repeaters is now becoming a holdup condition. The reason for this is due to their major source of energy dissipation. Thus, it is suggested by researchers to insert an optimum count of buffer circuits as repeaters inbetween long interconnect wires. This optimal number is judicially decided by doing calculations discussed in detail at Section 3.3 of this chapter so as to have a trade-off between power and delay. The various repeater design criteria, as illustrated in Figure 3.3, are selected based on the requirement to ensure a proper balance between power specs and delay values.

The total wire delay with repeaters is computed from Eqs. 3.01–3.10. Solving differential equation analytical formulas for optimal buffer size and the total optimal count are discussed in the aforementioned section. To enhance the electrical performance, the repeater buffer circuits are usually employed in lengthy interconnect wires. Figure 3.3 shows an interconnect wire with interpolation of equispaced repeater buffer. The interconnect wire is divided into k segments with each segment having length of lk. The buffer repeaters are h times the minimum size. The 50% propagation delay time can be computed from the equations under various scenarios. In addition to this, as predicted by various researchers, with a reduction in size of repeater and at the same time an increase in inter repeater length of interconnect wire, for a specific wire length the intrinsic buffer power consumption is suppressed while switching power because of total wire capacitance remains unaffected. Furthermore, a total count of repeater buffers embedded along the wire also is reduced; hence, it leads to great savings in terms of leakage power consumption. But the short circuit power may increase if the rise time increases.

Tables 3.16 and 3.17 show the impact of repeater buffer insertion at different wire lengths for three types of FETs (CMOSFET, CNTFET and GNRFET). The investigations for performance metrics like power and delay metrics are carried out at different buffer intervals interconnecting wire lengths ranging from 200 μm to 1,000 μm.

The result of Table 3.16 explicitly proves that the alteration in number of repeaters as well as variation in interconnect lengths both cause changes in value of impedance parameters. The average power dissipation is due to a combined effect of dynamic, leakage and short circuit power. The dynamic power depends on switching of transistor and hence measured by doing transient analysis. For the transient analysis a pulse width of 5 ns with a period of 10 ns is applied. The pulse has a rise and fall time of 10 ps each and simulation time of 100 ns is taken. From the tabulated data it is inferred that as the interconnect wire length increase then correspondingly power consumption also rises for the same number of repeater buffers. The capacitance of wire plays a key role in determining the average power consumption. As the GNRFET capacitance is least among all the different kinds of

TABLE 3.16
Estimation of average power dissipation with change in number of buffers at different wire lengths for CMOSFET, CNTFET and GNRFET paired with Cu interconnect

Length (μm)	No. of buffers	CMOSFET	CNTFET	GNRFET
200	2	7.42E-09	2.73E-10	1.7E-11
	6	2.507E-08	0.933E-09	2.72E-11
	12	3.604E-08	0.856E-09	3.40E-11
600	2	0.198 E-06	0.956 E-07	2.564 E-10
	6	2.625 E-06	0.956 E-07	22.03 E-10
	12	1.567 E-06	1.08 E-07	9.97E-10
1,000	2	14.734E-06	17.22E-07	74.09E-10
	6	59.21E-06	49.16E-07	125.56E-10
	12	106.46E-06	96.46E-07	164.99E-10

TABLE 3.17
Estimation of propagation delay with change in number of buffers at different wire lengths for CMOSFET, CNTFET and GNRFET repeater buffers paired with Cu interconnect

Length (μm)	No. of buffers	CMOSFET	CNTFET	GNRFET
200	2	1.979E-11	9.56E-10	3.89E-12
	6	1.18E-11	8.37E-10	2.46E-12
	12	1.05E-11	6.62E-10	2.05E-12
600	2	1.76E-09	44.96E-08	7.60E-10
	6	2.31 E-10	92.67 E-09	4.57E-11
	12	1.44E-11	9.65E-10	2.97E-12
1,000	2	32.06E-10	97.79E-09	29.15E-11
	6	11.55E-10	75.45E-09	6.56E-11
	12	8.36E-11	65.34E-10	63.05E-12

wires, therefore the average power dissipation is least for local and intermediate levels of interconnect wire. From the past studies, it is proved that as the carbon-based interconnects possess smaller optimal number of buffers than Cu counterparts, hence, pairing the CNTFET/GNRFET buffers with state-of-the-art interconnects results in a smaller power consumption. But in either cases with advancement of technology node as we move from 180 nm to 45 nm regime, there is significant rise in optimal number of repeater buffer count. Hence, it is a critical issue that the power dissipation becomes more prominent in advanced technology nodes of deep sub-micron technology and therefore requires acute attention.

Propagation delay is an important parameter of thought that measures the performance of ASIC. Repeater insertion is the best technique to mitigate the delay. Table 3.17 shows the effects of adding repeaters inbetween interconnect wire lengths. It can be inferred from Table 3.17 that the impact of increase in number of inserted repeaters is a reduction in overall delay for the conventional CMOSFETs as well as CNTFETs and GNRFETs. Besides this, it is also noted from the obtained results that with augmentation of more repeaters the difference between delay for CNTFET/GNRFET and CMOSFET keeps narrowing.

It has been observed from various literatures that different buffers designed using carbon nanotube FETs or GNRFETs show minimal delay and better performance than their counterpart Si CMOSFET. Moreover, instead of using Cu if carbon-based interconnects wires are used, it is observed that at the local level CNT/GNR-based interconnects have a marginal increase in value compared to their Cu counterpart because of the reason that at the local level ($l \leq \lambda$) and thus CNTs and GNRs operate in the ballistic region and have a large value of length independent intrinsic parasitic impedance associated with them, whereas for Cu, impedance changes with length. In contrast to this for global interconnects due to surface electron scattering and enhanced grain boundary scattering in addition to the presence of low conductive diffusion barrier layer leads to abrupt increase in resistivity of conventional Cu interconnects (if dimensions are of the order of electron mean free path in Cu). In general, the common trend observed is that the effective value of resistance and capacitance decreases with more repeaters for both types of wire material at various interconnect levels (intermediate, local and global) arranged in rising order of wire length.

Many researchers has proven that for a given delay penalty, the relative energy saving increases as the technology scales. This is primarily because of the fact that leakage power dissipation becomes the dominating component of the total power dissipation, and therefore reducing the repeater size and the number of repeaters results in large power savings. From Section 3.3, it is obvious that to achieve optimal power dissipation at a given delay penalty, the repeater size needs to be reduced and the interconnect length between successive repeaters needs to be increased. The total power savings increase as the technology scales. As illustrated in Figure 3.3(a), a typical minimum-sized VLSI gate will not be able to directly drive this DIL structure while still meeting the delay constraint. Henceforth, intermediate inverters are required to be introduced between the minimum sized gate and the interconnect repeater buffer. The ratio of the sizes of successive inverters is typically of the order of four so as to suppress the propagation delay. Thus, it can be finally concluded that on carrying out a detailed study to observe the influence of interconnect length on propagation delay, there is significant time delay diminution on increasing the number of buffer repeaters and the effect of excitation magnitude on time delay with a power delay product. It is observed that by using CNT interconnects paired with CNTFET or GNRFET, the total number of buffer repeaters and the time delay are suppressed by more than 40% and 50%, respectively, compared with the copper (Cu) interconnects.

3.8 SUMMARY

- This chapter gives an insightful explanation on the adoption of concept of repeater insertion methodology in VLSI on-chip interconnects for the purpose of signal restoration and delay reduction.
- It highlights the features of novel technology used for designing buffers to overcome the limitations encountered by conventional CMOS buffer insertion technology in terms of area and power consumption.
- It also discusses the idea of optimum number of buffers for achieving better performance.
- Critical issues regarding buffer size and placement strategies are also dealt with under the umbrella of state-of-the-art material used for modeling efficient nano interconnects integrated with CNTFET/GNRFET buffers.
- It is found that these carbon-transistors have reduced drain-induced-barrier-lowering (DIBL), better sub-threshold swing (SS) and higher conductivity than a Si CMOSFET.

- The simulation results obtained point out that low-dimensional carbon devices outperform silicon MOSFET in terms of power delay product (PDP) and energy delay product (EDP) by several orders of magnitude.
- For the equivalent amount of drain current, the channel area consumed by CNTFET and GNRFET is relatively less than for Si CMOSFET. This allows the fabrication of high-density ICs made of nanotubes and nanoribbons on chip.
- A key advantage of carbon-based logic gates over the Si CMOSFET counterparts is their low-energy consumption per logic transition.
- It is demonstrated through simulation that the CNTFET and GNRFET logic buffer repeater gates can have wiring lengths up to 5 μm before their performance becomes no better than a Si MOSFET.

EXERCISES

MULTIPLE-CHOICE QUESTIONS

Q1 Graphene is a zero bandgap semiconductor
 a True
 b False

Q2 CNTs stand for ___________.
 a Carbon nanotubes
 b Carbon nanotechnology
 c Carbon nanoscience and technology
 d Carbon nine technology

Q3 The carbon tubes have high conductivity.
 a True
 b False

Q4 If dimer lines (N) are N = 3p and N = 3p+1, then the
 a Bandgap is finite and the nanoribbons of graphene exhibits semiconducting characteristics.
 b Bandgap is very minute and the nanoribbons of graphene exhibits metallic characteristics.
 c Both a and b are true.
 d Both a and b are false.

Q5 Delay of buffers_____ with the increase in the power supply voltage.
 a Increase
 b Decrease
 c Remain constantdNone of these.

Q6 Delays of buffer _____ with the increase in the operating temperature.
 a Increase
 b Decrease
 c Remain constant
 d None of these.

Q7 The interconnect delay is
a Always less than the gate delay.
b Always more than the gate delay.
c Always equal to the gate delay.
d Can be more or less than the gate delay, depending on the technology node used.

Q8 Dynamic power dissipation of a CMOS inverter depends on the
a Power supply voltage.
b Channel width of the FET.
c Both a and b are true.
d Both a and b are false.

Q9 The goals of a repeater buffer inserted on chip nanointerconnect modeling is
a Minimization of area, increase operating speed and reduce power consumption.
bMinimize power consumption, decrease propagation delay and reduce area.
cMinimize power consumption, increase propagation delay and reduce area.
dMinimization of area, reduces operating speed and reduces power consumption.

Q10 The critical path in a design refers to
a The path having maximum delay.
b A path with minimum delay.
c The path having optimum delay.
d A path with no delay.

Q11 Lowest propagation delay through a buffer is due to
a Strong transistor, low temperature and high voltage.
b Weak transistor, high temperature and high voltage.
c Strong transistor, high temperature and high voltage.
d Weak transistor, low temperature and low voltage.

Q12 Dynamic power does not depend on __________.
a Transistor dimensions
b Load capacitance
c Power supply voltagedSwitching activity.

Q13 Short-circuit power dissipation does not depend on __________.
a Power supply
b Load capacitance
c Rise/fall time
d Transistor dimensions

Q14 Leakage power is due to __________.
a Sub-threshold current
b Leakage current
c Both a and b.
d None of these.

Q15 Battery-operated devices must have _________ power dissipation.
   a More
   b Less
   c Zero
   d None of these.

Q16 Low V_t transistors consume
   a More power.
   b Less power.
   c No power.
   d None of these.

Q17 High V_t transistors have
   a More speed.
   b Less speed.
   c No speed.
   d All of these.

Q18 Static power is more for
   a Slow input rise/fall time.
   b Fast input rise/fall time.
   c Slow output rise/fall time.
   d Fast output rise/fall time.

Short-Answer Questions

Q1 Differentiate between Schottky Barrier type field-effect transistor (SB-CNTFET) and MOS like CNTFET (MOS-CNTFET) with the help of an energy band diagram.
Q2 Define the term *threshold voltage* for CNT transistors and explain how the value of threshold voltage can be controlled by varying the diameter of nanotubes.
Q3 Differentiate between zigzag GNRs (ZGNRs) and armchair GNRs (AGNRs) with the help of a graphene lattice structure.
Q4 Write a short paragraph on process technology used for graphene fabrication.
Q5 What are interconnect delay models? Discuss them in detail. Also determine how propagation delay is computed by performing transient analysis.
Q6 Discuss how the leakage power is dissipated in a repeater inserted in an on-chip nano interconnect network.

Long-Answer Questions

Q1 Explain the mechanism of current transport in CNTFET and derive the equation for drain current and threshold voltage.
Q2 Derive the expression for dynamic power dissipation in an inverter. Explain how the switching activity affects the dynamic power dissipation.
Q3 Calculate the dynamic power dissipation of an inverter repeater buffer inserted interconnect network operating at a frequency of 200 MHz. The power supply voltage is 2.0 V and the load capacitance is 10 fF. If the propagation delay

through the repeater buffer is 10 ps, then calculate the value of the power delay product (PDP) and energy delay product (EDP).

Q4 What are the major sources of power dissipation in a repeater inserted in an on-chip nano-interconnect network?

Q5 Write a short paragraph on the power reduction techniques.

ANSWERS TO MULTIPLE-CHOICE QUESTIONS CHAPTER 3

1. A	10. A
2. A	11. A
3. A	12. A
4. A	13. D
5. B	14. C
6. A	15. B
7. D	16. A
8. A	17. B
9. B	18. A

REFERENCES

Akbari Eshkalak, Maedeh and Mohammad K. Anvarifard. "A Novel Graphene Nanoribbon FET with an Extra Peak Electric Field (EFP-GNRFET) for Enhancing the Electrical Performances." *Physics Letters* 381 (2017): 1379–1385.

Akinwande, Deji, Jiale Liang, Soogine Chang, Yoshio Nishi, and H.-S. Philip Wong. "Analytical Ballistic Theory of Carbon Nanotube Transistors: Experimental Validation, Device Physics, Parameter Extraction, and Performance Projections." *Journal of Applied Physics* 104 (2008): 124514.

Alpert, Charles J., Anirudh Devgan, and Stephen T. Quay. "Buffer Insertion for Noise and Delay Optimization." *IEEE Transactions on Computer-Aided Design Integra Circuits System* 18, no. 2 (1999): 1633.

Avouris, Phaedon, Joerg Appenzeller, Richard Martel, and Shalom J. Wind. "Carbon Nanotube Electronics." *Proceedings of IEEE* 91, no. 11 (2003): 1772–1784.

Bakoglu, H. B. and James D. Meindl. "Optimal interconnection circuits for VLSI." *IEEE Transactions on Electron Devices* ED-32, no. 5 (1985): 903–909.

Banerjee, Kaustav and Amit Mehrotra. "A Power Optimal Repeater Insertion Methodology for Global Interconnects in Nanometer Designs," *IEEE Transaction on Electron Devices* 49, no. 11 (2002): 2001.

Banerjee, Kaustav, Sheng-Chih Lin, Ali Keshavarzi, Siva Narendra, and V. De. "A self-Consistent Junction Temperature Estimation Methodology for Nanometer Scale ICs with Implications for Performance and Thermal Management," in Proceedings of International Electron Devices Meeting, Washington, DC, Dec. 2003, 887–890.

Chatzigeorgiou, Alexander, Spyridon Nikolaidis, and I. Tsoukalas. "Modeling CMOS Gates Driving RC Interconnect Loads." *IEEE Transactions on Circuits and Systems-Part II: Analog and Digital Signal Processing* 48, no. 4 (2001): 413–418.

Chau, Robert S., B. S., Doyle, M., Doczy, Suman, Datta, Scott A., Hareland, Ben, Jin, Jack, Kavalieros, and Matthew V., Metz. "Silicon Nano-Transistors and Breaking the 10 nm Physical Gate Length Barrier," *Device Research Conference* 2003 (2003): 123–126.

Chen, Guoqing and Eby G. Friedman. "Low Power Repeaters Driving RC and RLC Interconnects with Delay and Bandwidth Constraints," *IEEE Transactions on Very Large Scale Integration (VLSI) Systems* 14, no. 2 (February 2006): 161–172.

Chen, Jian Hao, Chaun Jang, Shudong Xiao, Masa Ishigami, and Michael S. Fuhrer. "Intrinsic and Extrinsic Performance Limits of Graphene Devices on SiO2." *Natural Nanotechnology* 3, no. 4 (2008): 206–209.

Chen, Ying-Yu, Artem Rogachev, Amit Sangai, Giuseppe Iannaccone, Gianluca Fiori, and Deming Chen, "A SPICE-Compatible Model of Graphene Nano-Ribbon Field-Effect Transistors Enabling Circuit-Level Delay and Power Analysis Under Process Variation." Design, Automation & Test in Europe Conference & Exhibition, 2013a, Grenoble, France.

Chen, Ying-Yu, Amit Sangai, Artem Rogachev, Mortez. Gholipour, Giuseppe Iannaccone, Gianluca Fiori, and Deming Chen. "A SPICE-Compatible Model of MOS-Type Graphene Nano-Ribbon Field-Effect Transistors Enabling Gate- and Circuit-Level Delay and Power Analysis Under Process Variation." *IEEE Transactions on Nanotechnology* 14, no. 6 (2015): 1068–1082.

Chen, Ying-Yu, Amit Sangai, Mortez Gholipour, and Deming Chen. "Schottky-Barrier-Type Graphene Nano-Ribbon Field-Effect Transistors: A Study on Compact Modeling, Process Variation, and Circuit Performance." *NanoArch* 2013b.

Datta, Suman, Tim, Ashley, J., Brask, Louise, Buckle, M., Doczy, Martin T., Emeny, David G., Hayes, Keith P., Hilton, R., Jefferies, Trevor, Martin, T. J., Phillips, David J., Wallis, P. J., Wilding, and Robert S., Chau. "85 nm Gate Length Enhancement and Depletion Mode InSb Quantum Well Transistors for Ultra High Speed and Very Low Power Digital Logic Applications." *IEEE International Electron Devices Meeting Technical Digest* (2005): 763–766.

Deng, Jie and H. S. Philip Wong. "A Compact SPICE model for Carbon-nanotube Field-effect Transistors including Nonidealities and Its Application-Part II: Full Device Model and Circuit erformance Benchmarking." *IEEE Transactions on Electron Devices* 54 (2007): 3195.

Elmore, William Cronk. "The Transient Response of Damped Linear Networks with Particular Regard to Wideband Amplifiers." *Journal of Applied Physics* 19, no. 1 (1948): 55–63.

Eshkalak, Maedeh Akbari, Rahim Faez, and Saeed Haji-Nasiri. "A Novel Graphene Nanoribbon Field Effect Transistor with Two Different Gate Insulators." *Physica E* 66 (2015):133–139.

Glasser, Lance A. and Daniel W. Dobberpuhl. *The Design and Analysis of VLSI Circuits.*Reading, MA: Addison Wesley, 1985.

Gonzalez, Ricardo E. and Mark A. Horowitz. "Energy Dissipation in General Purpose Microprocessors." *Journal Solid State Circuits* 31, no. 9 (1996): 1277–1284.

Han, Melinda Y., Barbaros Ozyilmaz, Yuanbo Zhang, and Philip Kim. "Energy Band-Gap Engineering of Graphene Nanoribbons." *Physical Review Letter* 98, no. 20 (2007): 206805.

Hanchate, Narender and Nagarajan Ranganathan. "Lector: A Technique for Leakage Reduction in CMOS Circuits." *IEEE Transactions on Very Large Scale Integration (VLSI) System* 12, no. 2 (February 2004): 196–205.

Illinois University GNRFET model website. Illinois University, Available: http://dchen.ece.illinois.edu/tools.html

Javey, Ali, Jing Guo, Damon B. Farmer, Qian Wang, E. Yenilmez, R. G. Gordon, M. Lundstrom, and H. Dai. "Self-Aligned Ballistic Molecular Transistors and Electrically Parallel Nanotube Arrays." *Nano Letters* 4 (2004a): 1319–1322.

Javey, Ali, Jing Guo, Damon B. Farmer, Qian Wang, Dunwei Wang, Roy G. Gordon, M. Lundstrom, and Hongjie Dai. "Carbon Nanotube Field-Effect Transistors with Integrated Ohmic Contacts and High-K Gate Dielectrics." *Nano Letters* 4 (2004b): 447–450.

Javey, Ali, Ryan Tu, Damon B. Farmer, Jing Guo, Roy. G. Gordon, and Hongjie Dai. "High Performance n-Type Carbon Nanotube Field-Effect Transistors with Chemically Doped Contacts." *Nano Letters* 5 (2005): 345–348.

Khursheed, Afreen, Kavita Khare, and Fozia Z. Haque. "Designing High-Performance Thermally Stable Repeaters for Nano-interconnects." *Journal of Computational Electronics* 18, no. 1 (2019): 53–64. DOI:10.1007/s10825-018-1271-0.

Kim, Dae-Hyun and Jesus A. del Alamo. "Logic Performance of 40 nm InAs HEMTs." (2007). International Electron Devices Meeting Technical DigestWashington, DC, 10–12 Dec. 2007.

Lin, Yu Ming, Christos Dimitrakopoulos, K. A. Jenkins, Damon B. Farmer, Hsin Ying Chiu, Alfred Grill, and Phaedon Avouris. "100-GHz Transistors from Wafer-Scale Epitaxial Graphene." *Science* 327 (2010): 662–662.

Lundstrom, Mark. Moore's Law Forever? *Science* 299 (2003): 210–211.

Magen, Nir, Avinoam, Kolodny, Uri, Weiser, and Nachum, Shamir. "Interconnect-Power Dissipation in a Microprocessor." *Proceedings of the ACM International Workshop on System Level Interconnect Prediction* 7–13, February 2004.

Naderi, Ali. "Theoretical Analysis of a Novel Dual Gate Metal–Graphene Nanoribbon Field Effect Transistor." *Materials Science in Semiconductor Processing* 31 (2015): 223–228.

Nougaret, Laurianne, Hugh Happy, Gilles Dambrine, Vincent Derycke, Jean-Philippe Bourgoin, Alexander Green, and Mark C. Hersam. "80 GHz field-Effect Transistors Produced Using High Purity Semiconducting Single-Walled Carbon Nanotubes." *Applied Physics Letters* 94 (2009): 243505.

Rabaey, Jan M. and Massoud Pedram. *Low Power Design Methodology*. Vol. 1. Kluwer Academic Publishers, 1996.

Raychowdhury, A., S. Mukhopadhyay, and K. Roy. "A Circuit-Compatible Model of Ballistic Carbon Nanotube Field-Effect Transistors." *IEEE Transactions on Computer-Aided Design of Integrated Circuits and Systems* 23, no. 10 (2004): 1411–1420.

Sandeep, Saini; Sreehari Veeramachanen; A. Mahesh Kumar; M. B. Srinivas. Schmitt trigger as an alternative to buffer insertion for delay and power reduction in VLSI interconnects. *IEEE Region 10 Conf TENCON* 2009 (2009): 1–5.

Saito, Riichiro, G. Dresselhaus, and Mildred S. Dresselhaus. "Physical Properties of Carbon Nanotubes," London: Imperial College Press, 1998.

Saxena, Prashant, Noel Menezes, P. Cocchini, and Desmond Kirkpatrick. "Repeater Scaling and Its Impact on CAD," *IEEE Transactions on Computer-Aided Design* 23, no. 4 (2004): 451–463

Secareanu, Radu M., Friedman Eby G. "Transparent repeaters," *ACM Great Lakes Symposium on VLSI* (2000): 63-66.

Shigyo, Naoyuki. "Tradeoff Between Interconnect Capacitance and RC Delay Variations Induced by Process Fluctuations." *IEEE Transactions on Electron Devices* 47 no.9 (2000): 1740–1744.

Son, Young-Woo, Marvin L. Cohen, and Steven G. Louie. "Energy Gaps in Graphene Nanoribbons," *Physical Review Letter* 2006.

Tans, Sander J., Alwin R. M., Verschueren, and C. Dekker. "Room-Temperature Transistor Based on Asingle Carbon Nanotube," *Nature* 393 (6680): 49–52, 1998

Thorat A., Khule R. S., Pable S. D. Design of High Speed Subthreshold Interconnects Using MCML Technique, August, 2016.

Venkatesan, R., J. A. Davis, K. A. Bowman and J. D. Meindl. "Optimal Repeater Insertion for N-Tier Multilevel Interconnect Architectures," *Proceedings of International Interconnect Technology Conference (IITC)* 2000, 132–134.

Venkatesan, Raguraman, Jeffrey A. Davis, and James D. Meindl. "Compact Distributed RLC Interconnect Models-Part IV: Unified Models for Time Delay, Crosstalk, and Repeater Insertion." *IEEE Transactions on Electron Devices* 50 4 (2003): 1094–1102.

Wang, Xinran, Yijian Ouyang, Xiaolin Li, Hailiang Wang, Jing Guo, and Hongjie Dai, "Room-Temperature All-Semiconducting Sub-10-nm Graphene Nanoribbon Field-Effect Transistors," *Physical Review Letters* 100 no. 20 (2008): 206803–206907.

Wessely, Pia Juliane, Frank, Wessely, Emrah, Birinci, Udo, Schwalke, and Bernadette, Riedinger. "Transfer-free Fabrication of Graphene Transistors." *Journal of Vacuum Science & Technology B, Nanotechnology and Microelectronics: Materials, Processing, Measurement, and Phenomena* 30, 2012. 10.1116/1.4711128

Wong, Philip H.-S., J. Appenzeller, Vincent Derycke, Richard. Martel, S. Wind, Phaedon H. Avouris. "Carbon Nanotube Field Effect Transistors – Fabrication, Device Physics, and Circuit Implications", Proceedings of International Solid State Circuits Conference, San Francisco, CA, Feb. 2003, 370–371, 2003.

Wong, Philip. H.-S. "Beyond the Conventional Transistor," *Solid-State Electron* 49 (2005): 755–762.

Yoon, Youngki, Gianluca Fiori, Seokmin Hong, Giuseppe Iannaccone, and Jing Guo. "Performance Comparison of Graphene Nanoribbon FETs with Schottky Contacts and Oped Reservoirs,"*IEEE Transactions on Electron Devices* 55, no. 9 (2008): 2314–2323.

4 Signal Integrity Analysis

4.1 INTRODUCTION

As VLSI technology begins to make headway into the DSM range, layout designs in terms of interconnects that satisfy timing and physical design constraints becomes intricate and trivial. The resemblance between a circuit "as fabricated" (on the wafer) and "as designed" (in the layout tool) grows ever weaker. Modeling on chip interconnects at 32 nanometer and smaller process technologies engender daunting challenges for the design engineers. Signal Integrity is an important figure of merit for high-speed, emerging on-chip interconnects. Past history reveals that the rapid rise in the rate of silicon failures due to signal integrity is mainly because of the inefficiency of existing design tools and methodologies.

The signal could be defined as information in the form of either a wave or a pulse, which is to be communicated between any two nodes through the interconnect wires. The term *integrity* purports to something that is unimpaired or intact in its actual form for a long time without distortion. Thus, together, the signal integrity could be defined as recapitulation of an entire signal while transmitting information from one node to another without any contortion in signal quality. In broader context to on-chip interconnects structure, the term *signal integrity* can be stated as the capability of an electrical signal to reliably carry information and defy the effect of high frequency electromagnetic interference from nearby signals.

In ultra deep-submicron designs, *timing* is dominated by an interconnect-dependent RC delay. Earlier repudiated as secondary effects, cross-coupling in between interconnect lines via and wire resistances and inductances, power integrity and interconnect wire self-heating has now eventually become first-order design parameters. ASIC design flows by means of a variety of existing point tools are unsuccessful to envisage final timing during early stages of the design. The growing complexity of system-on-chip design, coupled with uncorrelated tool flows, makes it more difficult to achieve design closure on all fronts. Henceforth, to ensure reliability and a successful tapeout, issues related to signal integrity must be resolved during the design flow.

Signal Integrity (SI) addresses two main issues of the nano interconnect design – the timing of the signal and its quality. The goal of signal integrity analysis is to ensure reliable high-speed data transmission through chip interconnects.

4.2 SIGNAL INTEGRITY: A CHALLENGE IN INTERCONNECT MODELING

The scaling of interconnected horizontal dimensions causes reduction in the aspect ratio of its horizontal to vertical dimensions. This leads to an increase in coupling

DOI: 10.1201/9781003104193-4

capacitance to ground capacitance ratios of interconnected wire. Determined by the amount of mutual capacitance and relative rate of signal switching, there is a significant influence of crosstalk noise on the interconnect wire. In addition to this at high operating frequencies, the tightly packed interconnects can introduce transient crosstalk noise. This crosstalk noise is conditioned on its amplitude and when it happens, can initiate false switching or signal propagation delay uncertainty on the victim net, thereby resulting in circuit malfunction or system failure.

Thus, crosstalk can be defined as the undesired effect created on one interconnected line due to a signal transmitted on another interconnected line. These switching interconnected line are termed a victim net and aggressor net, respectively. Contingent on the pattern of switching transitions in the coupled interconnect lines, crosstalk is categorized as functional and dynamic crosstalk. If the victim net is inert (not switching), a voltage spike appearing on it due to switching in an adjoining interconnected line (termed *aggressor*) is referred as the functional crosstalk. If both of the adjoining interconnected lines (victim as well as aggressor) are simultaneously switching either in-phase or out-phase then such type of crosstalk is called dynamic crosstalk. The succeeding sections discuss in detail how change in logic value and propagation delay occurs under functional and dynamic crosstalk, respectively. Furthermore, the crosstalk noise causes signal overshoot, undershoot and ringing effects. Therefore, accurate estimation of performance parameters, under the influence of crosstalk, becomes a prime goal for designing high-performance on-chip interconnects.

4.3 CROSSTALK MECHANISM

The phenomenon of crosstalk happens by means of following two mechanisms:

- Inductive crosstalk: caused by magnetic field
- Electrostatic crosstalk: caused by electric field

Inductive Crosstalk

The phenomenon of inductive crosstalk is described with the help of Figure 4.1.

As shown in Figure 4.1, there are two nets: NET1 and NET2. NET1 is connected to signal source V_s and carries a current I_s while the NET2 is quiescent and is connected to the ground via resistance R. Any variation in signal source voltage V_s causes change in current I_s flowing through NET1. This variation in current I_s creates a magnetic field around it and a mutual inductance will exist between these two nets, as shown in Figure 4.1. An effect of mutual inductance M, a voltage V_n across resistance R will develop. Hence, the equivalent circuit in Figure 4.1 shows electrical analysis.

Electrostatic Crosstalk

The phenomenon of electrostatic crosstalk is described with the help of Figure 4.2.

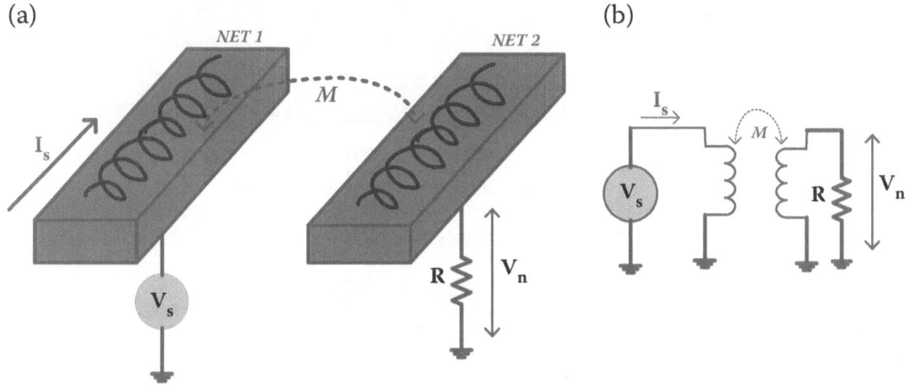

FIGURE 4.1 Inductive crosstalk.

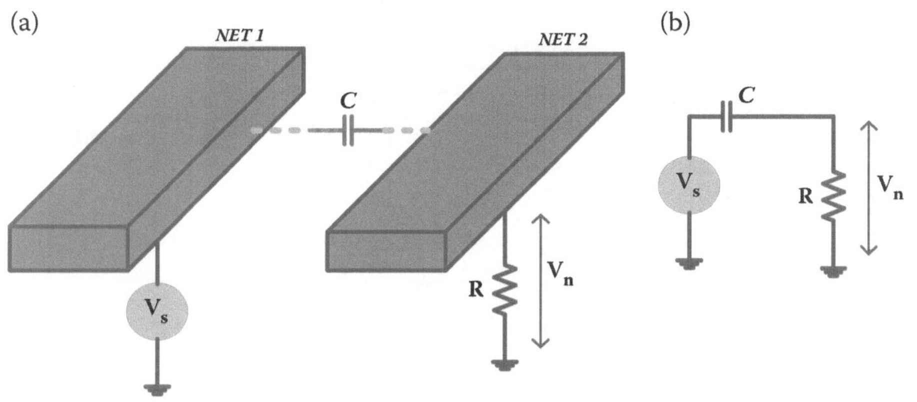

FIGURE 4.2 Electrostatic crosstalk.

The voltage connected to NET1 creates an electric field around it. If there is any variation in the electric field, then it can either radiate radio waves or can capacitive couple to adjoining nets. Such coupling of the electric field is called electrostatic coupling.

Out of the two crosstalk mechanisms discussed, the electrostatic crosstalk is more problematic and significant than inductive crosstalk.

The main cause of electrostatic crosstalk is the coupling capacitance in between interconnect nets lying either in same layer or in different layers if it is a six or five metal process. As illustrated in Figure 4.3, there can be three types of coupling capacitances: C_s (substrate capacitance: between M1 and substrate), C_{iL} (interlayer capacitance: between M1 and M2), C_i (lateral capacitance: between the same layers of metal).

Depending on the pattern in which switching of signals occurs in the aggressor net and victim net, there can be four cases. As illustrated in Figure 4.4, there is one aggressor net and one victim net connected as a DIL (driver-interconnect-load) test bench structure.

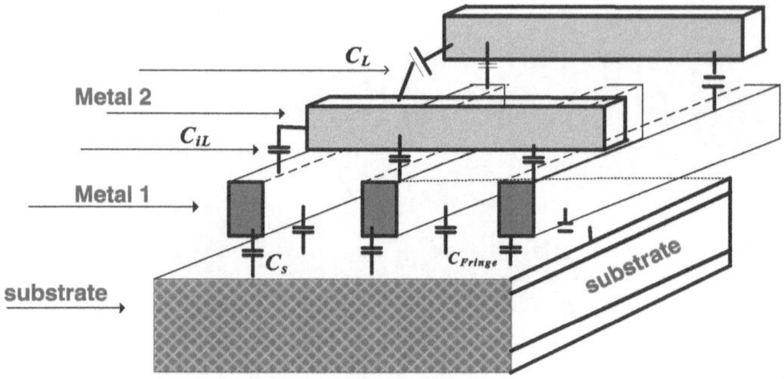

FIGURE 4.3 Types of capacitance allied with interconnect.

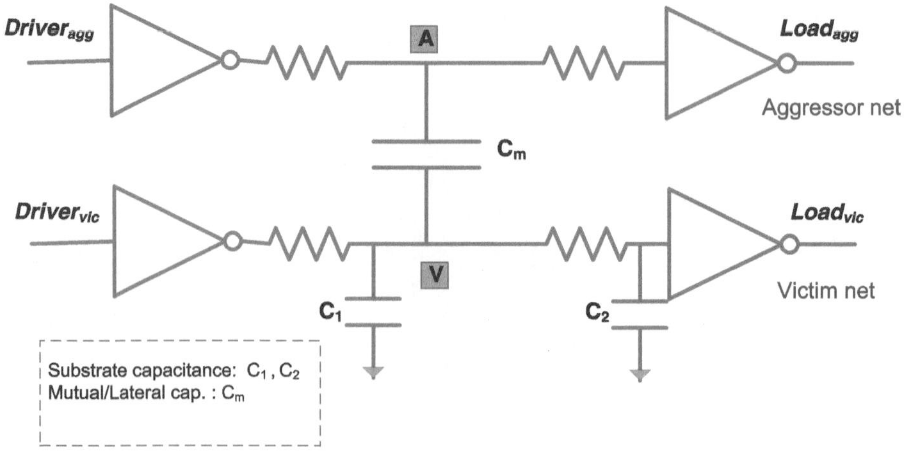

FIGURE 4.4 DIL structure of aggressor net coupled with victim net.

Case I: If *aggressor* input is switching from High to Low and *victim is quiescent* i.e. victim input is held constant at 1 (high steady state).
Case II: If *aggressor* input is switching from Low to High and *victim is quiescent* i.e. victim input is held constant at 0 (low steady state).
Case III: If *aggressor* input is switching from High to Low and *victim* input is switching from Low to High.
Case IV: If *aggressor* input is switching from High to Low and *victim* input is also switching from High to Low.

Crosstalk noise can be two types, namely, crosstalk glitch and crosstalk delay.

- ***Crosstalk Glitch:*** As mentioned in *Case I* and *Case II,* crosstalk glitch is caused on a logic level fixed victim signal due to the coupling of switching activity occurring on the adjoining aggressor net. Hence, because of some charge transferred by the switching aggressor nets by means of coupling

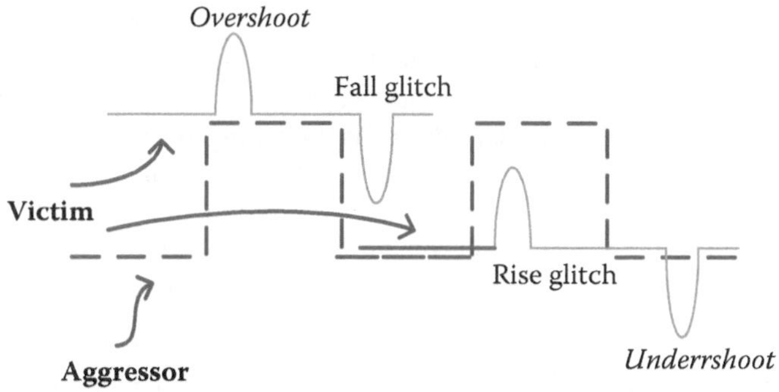

FIGURE 4.5 Crosstalk glitch.

capacitances onto the steady-state victim net results in *crosstalk glitch.* The magnitude of the glitch determines if it can result into a functional failure. It can be categorized as (i) rise/fall glitch and (ii) overshoot/undershoot glitch, as shown in Figure 4.5.

i. Rise and fall glitch: Aggressor net with a rising input pulse instigates a rise glitch on a *quiescent* victim net whose input is held constant at the logic 0 level, while an aggressor net with a falling input pulse instigates a fall glitch on a quiescent victim net whose input is held constant at logic 1 level.
ii. Overshoot and undershoot glitch: Aggressor net with a rising input pulse instigates an overshoot glitch on a victim net whose input is steady high. It triggers the victim net voltage above its steady high value. While an aggressor net with a falling input pulse instigate an undershoot glitch on a victim net whose input is steady low. It triggers the victim net voltage below its steady low value.

- ***Crosstalk Delay:*** As mentioned in *Case III* and *Case IV,* crosstalk delay is due to the coupling between the switching activity of the victim and the switching activity of the aggressors, which results in the alteration of timing on a victim net signal. Crosstalk delay is contingent on the direction of signal propagation in aggressor and victim nets. It causes either slower transition or faster transition of victim nets. It can be categorized as (i) negative crosstalk delay and (ii) positive crosstalk delay.

i. Negative crosstalk delay: In the case of when the aggressor and victim nets both are switching simultaneously in the same direction, it results in a smaller delay for the victim net. The fall in delay time is known as a negative crosstalk delay.
ii. Positive crosstalk delay: In the case of when the aggressor and victim nets are both switching simultaneously in the opposite direction, it results in a larger delay for the victim net. The rise in delay time is known as a positive crosstalk delay.

These are discussed in detail by analyzing each case separately.

Case I: If the aggressor input is switching from high to low and the victim is quiescent i.e. victim input is held constant at 1 (high steady state).

Here, if the input of the aggressor driver is switching from logic level high to a logic level low while the input of victim driver is held constant at logic 1 (high logic level), then the output of victim driver should ideally remain constant to zero. But in actuality as the input to aggressor driver switches from high to low, then the potential at node $\boxed{A}$ starts increasing from low to high. Due to a rise in potential at node $\boxed{A}$, the mutual capacitance C_m starts charging. This capacitance acts as leaky capacitance and hence some charge will transfer to node $\boxed{V}$, thereby causing a rise in the voltage at node $\boxed{V}$. Hence, the output level of victim driver instead of remaining constant at zero experiences a positive spike or rising glitch, as shown in the Figure 4.6. The magnitude of the glitch or intensity of the spike in victim driver output voltage depends on various factors discussed in later sections.

Case II: If the aggressor input is switching from low to high and the victim is quiescent i.e. victim input is held constant at 0 (low steady state).

In this case, the victim input is held constant at logic low (Level 0) and ideally victim output node $\boxed{V}$ should remain constant at Logic level 1. As the aggressor input switches from low to high, then correspondingly the aggressor output node $\boxed{A}$ will switch from high to low. Now, with the switching of node $\boxed{A}$ from high (Level 1) to low (Level 0) and node $\boxed{V}$ at Level 1, there will be a potential difference from node $\boxed{V}$ to node $\boxed{A}$. Due to mutual capacitance C_m, transfer of charge takes place from victim net towards aggressor net, which will lower the potential of node $\boxed{V}$. This decline in potential at node $\boxed{V}$ will result in negative/falling glitch, as seen in Figure 4.7.

For both of the cases, the victim net is quiet and its input is kept constant at a fixed voltage. Whenever there is a voltage transition in one or more coupled aggressor nets, then the sufferer or victim net will experience voltage perturbations, termed a *glitch*. In some cases, these spurious signals called glitches are so

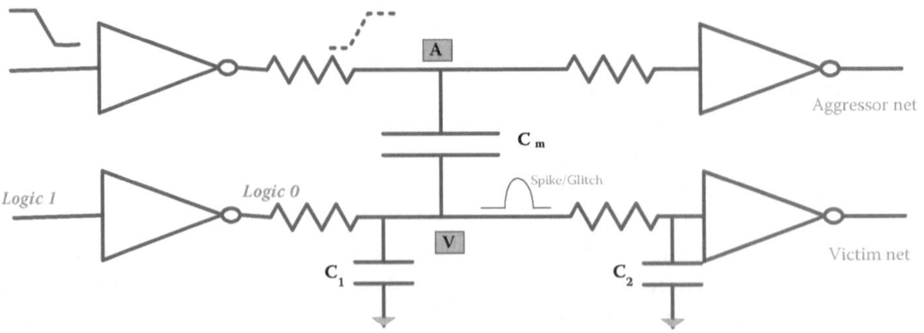

FIGURE 4.6 Aggressor switching from high to low and the victim is held constant at a high steady state.

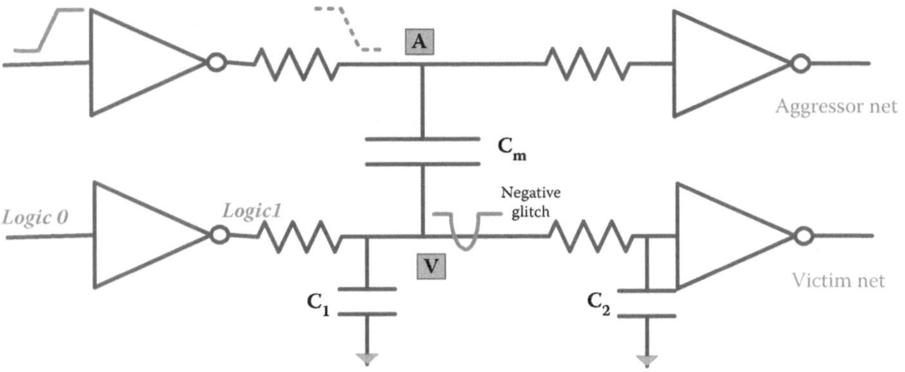

FIGURE 4.7 Aggressor switching low to high and victim is held constant at low steady state.

small that their effect on a circuit response is insignificant. In other cases, if the magnitude of the glitch is significant enough to affect the circuit's response, then it may cause functionality failure. Thus, it can be stated that the glitch could be either potentially unsafe and result in system failure or it may be safe; it all depends on the glitch height.

In the preceding section, analysis of how the voltage waveform characteristics of a spurious signal (glitch) depend on line and driver parameters, how this spurious signal propagates throughout the logic of the circuit and finally, how it can be measured is discussed.

Effect of Crosstalk Glitch

For a driver interconnect load structure, on the basis of the height and duration of the spurious signal, the next digital repeater gate might interpret it as an actual valid signal, thereby causing a logic error that can be further propagated. This error can be enduring if it affects the circuit's sequential block. In combinational blocks, only transient errors will appear. The main waveform parameters influencing the effect of spurious signals through logic are its amplitude and its width.

As shown in Figure 4.8, so as to cause a logic error in a digital repeater gate, the noise signal amplitude must be elevated than the logic threshold voltage (switching voltage) of the affected repeater gate (the gate with the glitch signal at its input). If the amplitude does not attain this value, a logic error cannot be generated, and for this case we do not consider spurious signal propagation through the gate and is in a safe region. In an undefined region when the amplitude is greater than the switching voltage of the affected repeater gate, then the possibility of propagation of the signal will depend on its width and also on the propagation delay of the affected repeater gate.

As the technology nodes shrinks and we lower the supply voltage, then the reduction in noise margin occurs, which further results in a high impact of crosstalk glitch.

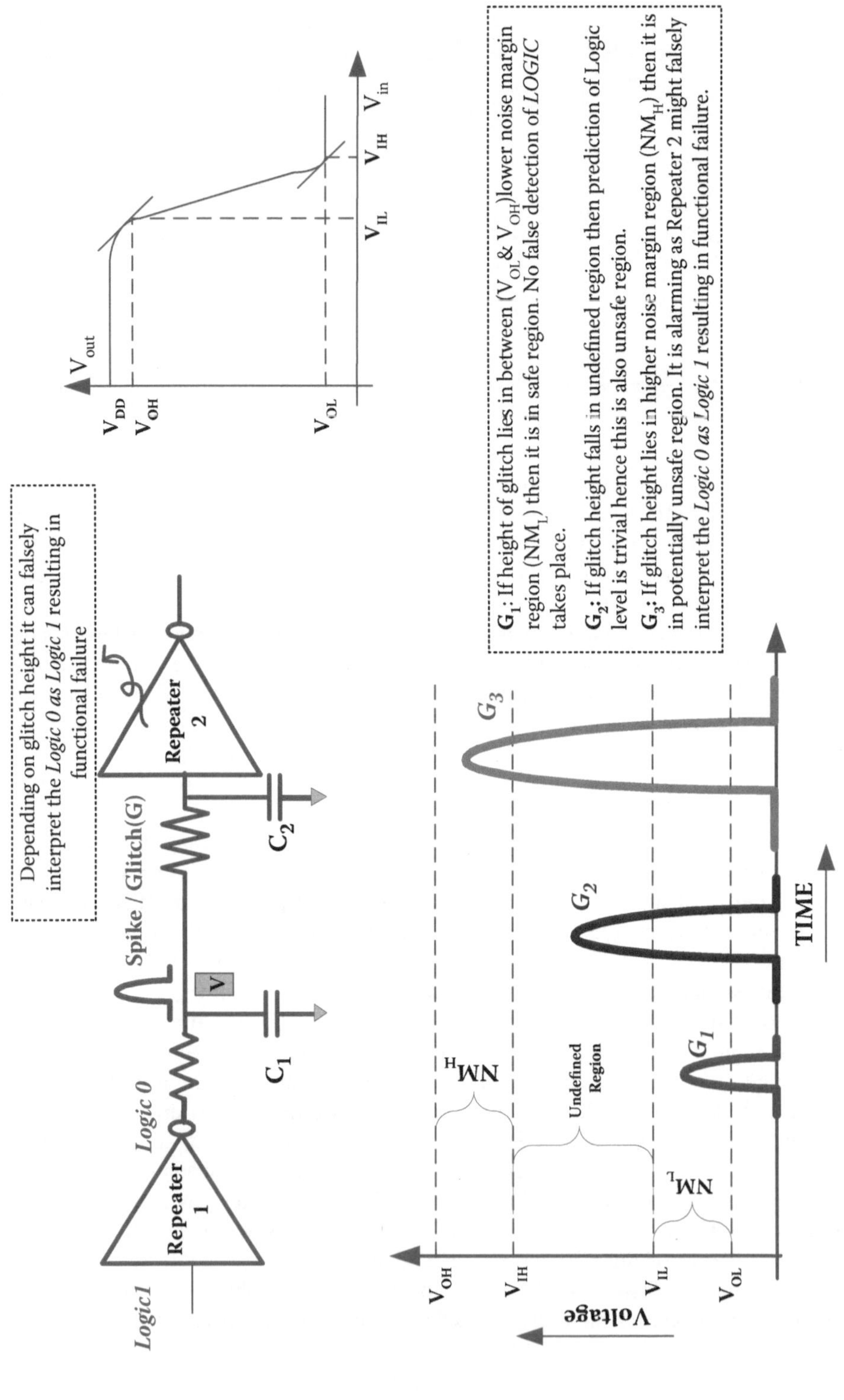

FIGURE 4.8 Effect of glitch height in context to noise margin.

Thus, a potentially unsafe glitch can be dangerous as it could result in false prediction of logic level, which may possibly set or reset the memory data and eventually causes functional failure. The penetration depth of a glitch signal is determined by the number of repeater gates it can actively transverse until a logical error is caused at the last one. Hence, the factor on which the amplitude height of a crosstalk glitch depends is discussed in a subsequent section.

Factors on Which Height of Crosstalk Depends

1. The coupling capacitance
2. The aggressor's driver strength
3. The victim's driver strength

The coupling capacitance (c_m) demonstrates the effect of crosstalk induced delay that primarily depends on the spacing (S_p) between aggressor and victim lines and can be expressed as

$$c_{cm} = \frac{\pi \varepsilon_0 \varepsilon_r}{\cosh^{-1}(S_p)} \tag{4.1}$$

Closer is the distance between aggressor and victim net, larger will be coupling capacitance and in turn larger will be crosstalk glitch height. The lower metal layers has minimum spacing hence they are more prone to crosstalk.

In addition to this the strength of aggressor driver and victim driver also demonstrates the effect of crosstalk. Higher the aggressor driver strength faster is the slew rate i.e switching is very fast; as a result higher will be the crosstalk glitch. Whereas higher the driver strength difficult it will be to change the output logic; as a result lower will be the crosstalk glitch.

Case III: If aggressor input is switching from high to low and victim input switching low to high.

In this case transition occurs in both aggressor and the victim net but in opposite direction. Here the aggressor input is switching high to low and simultaneously the victim input is switching low to high. Initially the output of victim driver net will transcend from high to low but as the potential of node $\boxed{A}$ is rising with time then because of coupling capacitance C_m there is transfer of charge from node $\boxed{A}$ to node $\boxed{V}$; which results in bump at the victim net $\boxed{V}$. This bump in the victim driver output waveform causes increase of slew rate. Greater the value of the coupling capacitor bigger will be the bump. This bump also affects static timing analysis of the circuit. Thus due to crosstalk the delay of victim cell has increased in this case. As delay of chip changes the latency also changes, leading to setup timing violation.

Case IV: If aggressor input is switching from high to low and victim input is switching high to low.

Here, both the aggressor and the victim nets are switching in the same direction. With the rise in potential at node $\boxed{A}$, there will be a transfer of some charge by the coupling capacitance C_m to node $\boxed{V}$. This coupling results in crosstalk, which causes a reduction in delay time. With a reduction in delay, the transition time becomes faster, which leads to hold time violation.

4.4 CROSSTALK ANALYSIS

The DIL setup (shown in Figures 4.7–4.10) with capacitive coupled aggressor and sufferer interconnect nets, mentioned previously is considered for carrying out the crosstalk analysis. The crosstalk propagation delay and crosstalk noise is analyzed for three different types of interconnect structures in this section.

For a very high frequency of operation, the tightly packed interconnects generates transient crosstalk. This crosstalk, as mentioned earlier, has a strong impact on signal propagation delay and in some cases might result in logic or functional failure. Hence, it is enviable to develop a perfect pedagogical model for analytically analyzing the crosstalk effects occurring in nano interconnects.

Past history reflects that numerous analytical models have been recommended by researchers (Kaushik and Sarkar, 2008; Rossi et al., 2007) in order to study the FET-gate-driven coupled interconnects nets. But unpropitious is that these models consider FET drivers as purely resistive linear load (Agarwal et al., 2006; Orlandi and Paul, 1996). However, in reality, these FET gates are driving the aggressor or victim nets operate in cutoff, linear and saturation regions during the input-output transition period. Thus, considering that during the transition state the driver FET solely operates in the linear region, this results in erroneous performance estimation of the DIL system. The main reason for this error is that in order to solve transmission lines with partial differential equations, analysis is carried out in a frequency domain while the analysis for the FET driver is carried out in a time domain. Hence, due to a conversion issue of the frequency/time domain, the discrepancy in results happens.

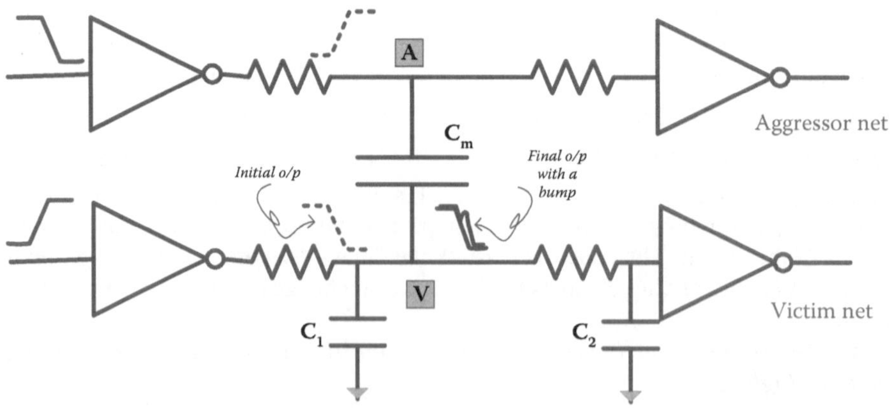

FIGURE 4.9 Aggressor switching high to low and victim input switching low to high.

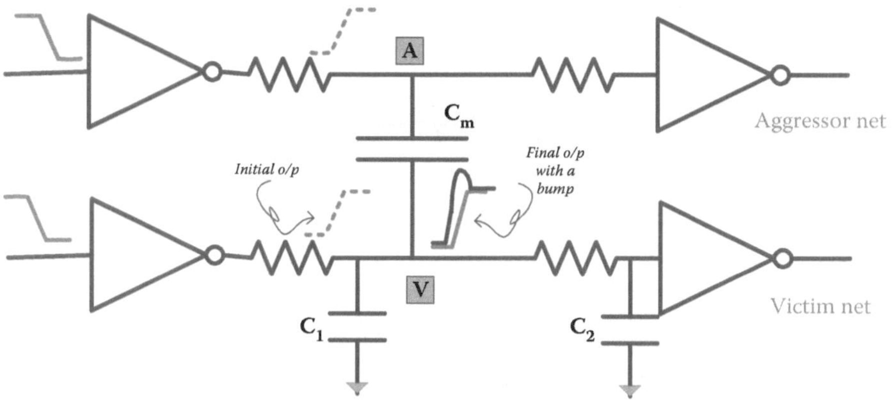

FIGURE 4.10 Aggressor switching high to low and victim input switching high to low.

Hence, bearing in mind the nonlinear effects of FET drivers, a precise and time-efficient model is discussed in this section to commensurate the crosstalk incited performance analysis of nano interconnects. The model discussed is formulated by a finite difference time domain (FDTD) methodology for the DIL system. The voltage and current values can be predicted at any specific point of interconnect net using FDTD methodology. Here the FET driver is modeled by the nth power law (Li et al., 2011) by taking into consideration a finite drain conductance parameter. HSPICE simulations validate the computational result obtained using this model.

Prolegomenon to FDTD Technique

Finite difference time domain (FDTD), a popularly known Yee technique, is widely used for the computation of electromagnetic problems. In this technique, using the central difference approximations, the time-dependent partial differential equations are discretized in space and time (Yee, 1966). It is used for modeling interconnects as transmission lines. For analyzing interconnects using FDTD, a computational domain is created. Interconnects as distributed RLC with relative parameters of V and I is taken. For the coupled interconnect nets, the mutual coupling capacitance has to be incorporated within the computational domain. Contained by the computational domain, the values of V and I are determined at each point in space and time. Once the computational domain is built, then near-end and far-end boundaries are determined to match the FDTD solution for voltage and current at discrete time and space points. Time is implicit in the FDTD method, whereas space is explicit.

4.4.1 FDTD Model for Crosstalk Analysis of FET-Driven Coupled Copper Interconnects

The FDTD technique is considered in this section for crosstalk analysis of two coupled Cu interconnects driven by an FET driver in a nonlinear mode (Kumar et al., 2014). It takes into account the nonlinear effects of an FET driver and is modeled using the nth power law model. In addition to this, the short-channel

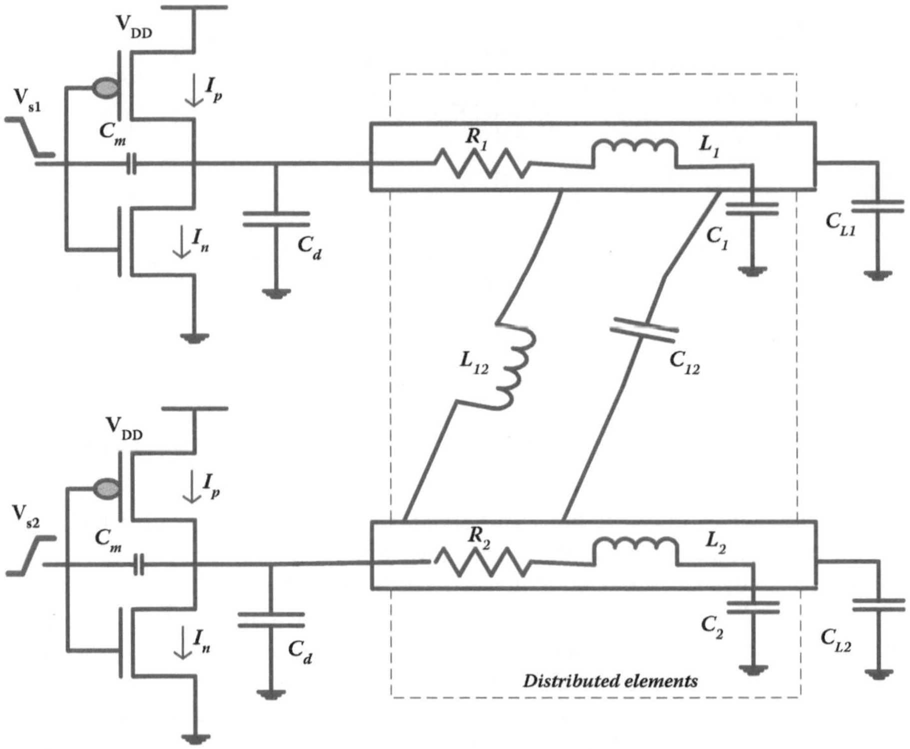

FIGURE 4.11 Coupled aggressor (Line 1) victim (Line 2) interconnect nets driven by FET gate.

effects, including the velocity saturation and finite drain conductance parameter, are also well thought out in the proposed model.

As illustrated in Figure 4.11, there are two coupled interconnect lines (aggressor and victim) driven by FET gates. The aggressor and victim nets are modeled using a distributed RLC parasitic. Δx denotes the minute distance. In the FDTD method, V and I variables are computed in time and space domains alternatively, as depicted in Figure 4.12. The time is denoted by t and space is denoted by x along the length of the interconnect net.

From the telegrapher's equations, the expression for coupled transmission line becomes

$$\frac{\partial}{\partial x} V_1(x, t) + L_1 \frac{\partial}{\partial t} I_1(x, t) + L_{12} \frac{\partial}{\partial t} I_2(x, t) + R_1 I_1(x, t) = 0 \tag{4.2}$$

$$\frac{\partial}{\partial x} V_2(x, t) + L_2 \frac{\partial}{\partial t} I_2(x, t) + L_{12} \frac{\partial}{\partial t} I_1(x, t) + R_2 I_2(x, t) = 0 \tag{4.3}$$

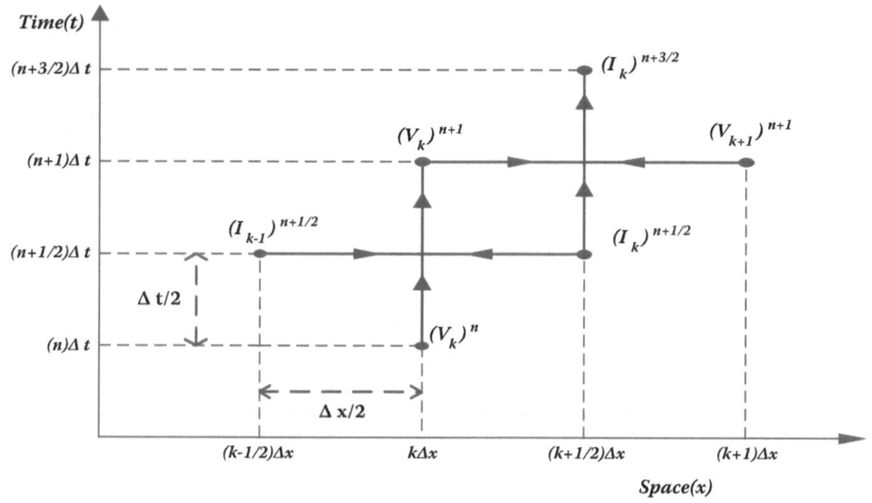

FIGURE 4.12 Discretized voltage and current along space and time.

$$\frac{\partial}{\partial x} I_1(x,t) + (C_1 + C_{12}) \frac{\partial}{\partial t} V_1(x,t) - C_{12} \frac{\partial}{\partial t} V_2(x,t) = 0 \tag{4.4}$$

$$\frac{\partial}{\partial x} I_2(x,t) + (C_2 + C_{12}) \frac{\partial}{\partial t} V_2(x,t) - C_{12} \frac{\partial}{\partial t} V_1(x,t) = 0 \tag{4.5}$$

In matrix form, the above four equations are given as

$$\frac{\mathrm{d}}{\mathrm{dx}} \boldsymbol{V}(x,t) + \boldsymbol{RI}(x,t) + \boldsymbol{L}\frac{\mathrm{d}}{\mathrm{dt}}\boldsymbol{I}(x,t) = 0 \tag{4.6}$$

$$\frac{\mathrm{d}}{\mathrm{dxz}} \boldsymbol{I}(x,t) + \frac{\mathrm{d}}{\mathrm{dt}}\boldsymbol{CV}(x,t) = 0 \tag{4.7}$$

$$\boldsymbol{V} = \begin{bmatrix} V_1 \\ V_2 \end{bmatrix}, \boldsymbol{I} = \begin{bmatrix} I_1 \\ I_2 \end{bmatrix}, = \begin{bmatrix} R_1 & 0 \\ 0 & R_2 \end{bmatrix}, \boldsymbol{L} = \begin{bmatrix} L_1 & L_{12} \\ L_{12} & L_2 \end{bmatrix} \text{ and } \boldsymbol{C} = \begin{bmatrix} C_1 + C_{12} & -C_{12} \\ -C_{12} & C_2 + C_{12} \end{bmatrix}$$

Interconnect net rail is decimated into N divisions for the implying FDTD. Initially, the number of points in space and time domains is defined while implementing the FDTD method. N and K are total points along time and space, respectively. Along with space and time domains, points are discretized as $k\Delta x$ and $n\Delta t$, respectively, where k and n are integers and defined as $1 \le k \le K$ and $1 \le n \le N$ (Agrawal and Chandel, 2012) (Figure 4.13).

With the application of finite central difference approximation; the equations obtained are

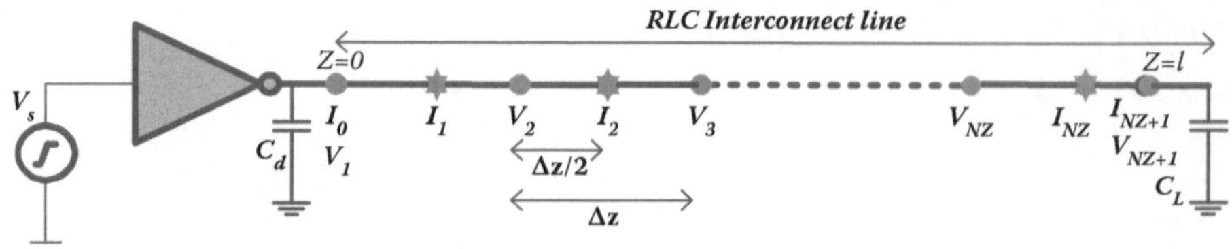

FIGURE 4.13 Space discretization of FDTD technique on Cu interconnects net.

$$\frac{V_{k+1}^{n+1} - V_k^{n+1}}{\Delta x} + L\frac{I_k^{n+3/2} - I_k^{n+1/2}}{\Delta t} + R\frac{I_k^{n+3/2} + I_k^{n+1/2}}{2} = 0 \tag{4.8}$$

$$I_k^{n+3/2} = BDI_k^{n+1/2} + B(V_k^{n+1} - V_{k+1}^{n+1})\ for\ k = 1, 2, \ldots. N_x \tag{4.9}$$

where $= \left[\frac{\Delta x}{\Delta t}L + \frac{\Delta x}{2}R\right]^{-1}$, $D = \left[\frac{\Delta x}{\Delta t}L - \frac{\Delta x}{2}R\right]$

$$\frac{I_k^{n+1/2} - I_{k-1}^{n+1/2}}{\Delta x} + C\frac{V_k^{n+1} - V_k^n}{\Delta t} = 0 \tag{4.10}$$

$$V_k^{n+1} = V_k^n + A(I_{k-1}^{n+1/2} - I_k^{n+1/2})\, for\ k = 2, 3, \ldots N_x \tag{4.11}$$

where $A = \left[\frac{\Delta x}{\Delta t}C\right]^{-1}$

With the application of these equations, voltage and current can be determined along the interconnect rail. To define the deriving source at the driver end and load at the far receiver end, near-end and far-end boundary conditions, respectively, need to be incorporated. These are described as follows.

Assimilation of Near End and Far End Boundary Conditions

a Near-End Boundary Condition

At the near end boundary conditions, put $k = 1$ in Eq. 4.11; then the expression becomes

$$V_1^{n+1} = V_1^n + 2A(I_0^{n+1/2} - I_1^{n+1/2}). \tag{4.12}$$

The source current I_0 at $(n + (1/2))$ time interval is computed by taking the average of values at n and $n + 1$; the expression becomes

$$V_1^{n+1} = V_1^n + 2A\left(\frac{I_0^{n+1} + I_0^n}{2} - I_1^{n+1/2}\right). \tag{4.13}$$

The I_0 denotes the driving current of FET. The expression for driving current is obtained by applying KCL at the near-end boundary condition:

$$I_0 = C_m\left[\frac{d(V_s - V_1)}{dt}\right] + I_p - I_n - C_d\frac{dV_1}{dt} \tag{4.14}$$

where C_m and C_d are the drain to gate coupling capacitance and drain diffusion capacitance, respectively. The current flowing in pFET is denoted by I_p and in nFET is denoted by I_n.

Applying the nth power law model (Sakurai and Newton, 1990), these currents can be expressed as

$$I_p = \begin{cases} 0; & V_s \geq V_{DD} - |V_{Tp}|\ (cutoff) \\ I_{DSATp\left(1+\sigma_p(V_{DD}-V_1)\right)}\left(2 - \frac{V_{DD}-V_1}{V_{DSATp}}\right)\frac{V_{DD}-V_1}{V_{DSATp}}; & V_1 > V_{DD} - V_{DSATp}\ (linear) \\ I_{DSATp(1+\sigma_p(V_{DD}-V_1))}; & V_1 \leq V_{DD} - V_{DSATp}\,(saturation) \end{cases}$$

$$I_n = \begin{cases} 0; & V_s \leq V_{Tn}\ (cutoff) \\ I_{DSATn(1+\sigma_n(V_1))}\left(2 - \frac{V_1}{V_{DSATn}}\right)\frac{V_1}{V_{DSATn}}; & V_1 < V_{DSATn}\ (linear) \\ I_{DSATn(1+\sigma_n(V_1))}; & V_1 \geq V_{DSATn}\ (saturation) \end{cases}$$

V_T and σ denotes threshold voltage and finite drain conductance parameter, whereas drain saturation voltage is denoted by V_{DSAT} and drain saturation current is given by I_{DSAT}.

$$V_{DSATp} = K_p(V_{DD} - V_s - |V_{Tp}|)^{m_p} \tag{4.15}$$

$$V_{DSATn} = K_n(V_s - V_{Tn})^{m_n} \tag{4.16}$$

$$I_{DSATp} = \frac{W_p}{L_{eff}}B_p(V_{DD} - V_s - |V_{Tp}|)^{s_p} \tag{4.17}$$

$$I_{DSATn} = \frac{W_n}{L_{eff}}B_n(V_s - V_{Tn})^{s_n} \tag{4.18}$$

Parameters s and B determine the saturation region, while m and K determine the linear region characteristics. W is the effective channel width and L is the channel length of FET.

The source current I_0 is discretized as:

$$I_0^{n+1} = C_m\frac{V_s^{n+1} - V_s^n}{\Delta t} + I_p^{n+1} - I_n^{n+1} - (C_m + C_d)\frac{V_1^{n+1} - V_1^n}{\Delta t} \tag{4.19}$$

$$V_1^{n+1} = V_1^n + \boldsymbol{EA}\left[\frac{\boldsymbol{C_m}}{\Delta t}(\boldsymbol{V}_s^{n+1} - \boldsymbol{V}_s^n) + \boldsymbol{I}_0^n\right] - 2\boldsymbol{EAI}_1^{n+1/2} + \boldsymbol{EA}(\boldsymbol{I}_p^{n+1} - \boldsymbol{I}_n^{n+1}) \tag{4.20}$$

$\boldsymbol{E} = \left[\boldsymbol{U} + \frac{A}{\Delta t}(C_m + C_d)\right]^{-1}$ and $\boldsymbol{U}$ is the identity matrix.

b Far-End Boundary Condition

At the far end, the boundary conditions put $k = Nx + 1$ in Eq. 4.11 and then the expression for voltage at the receiver end becomes

$$\boldsymbol{V}_{Nx+1}^{n+1} = \boldsymbol{V}_{Nx+1}^{n} + 2\boldsymbol{A}\left(\boldsymbol{I}_{Nx}^{n+1/2} - \frac{\boldsymbol{I}_{Nx+1}^{n+1} + \boldsymbol{I}_{Nx+1}^{n}}{2}\right). \tag{4.21}$$

The output current for the capacitive load C_L is given by

$$I_{Nx+1} = C_L \frac{d}{dt} V_{Nx+1} \tag{4.22}$$

The discretized value is

$$\boldsymbol{I}_{Nx+1}^{n+1} = C_L \frac{(\boldsymbol{V}_{Nx+1}^{n+1} - \boldsymbol{V}_{Nx+1}^{n})}{\Delta t} \tag{4.23}$$

$$\boldsymbol{V}_{Nx+1}^{n+1} = \boldsymbol{V}_{Nx+1}^{n} + 2\boldsymbol{FA}\left(\boldsymbol{I}_{Nx}^{n+1/2} - \frac{\boldsymbol{I}_{Nx+1}^{n}}{2}\right). \tag{4.24}$$

$$\boldsymbol{F} = \left[U + \frac{AC_L}{\Delta t}\right]^{-1}$$

The boundaries conditions are implicitly derived and interconnect equations are explicitly derived. Hence, there is no issue of stability at the boundaries and stability of the DIL system is solely determined by interconnect wire.

4.4.2 FDTD Model for Crosstalk Analysis of Carbon Nanotube (CNT) Interconnects

The assimilation of different boundary conditions in FDTD models is a tricky task. Initially boundary conditions were assimilated to analyze transmission lines for resistive driver and resistive load boundaries (Paul, 1994, 1996). But unpropitious is that all these models focused only on copper interconnects and may not be appropriate for state of art graphene-based nano interconnects. Incorporating the effect of quantum and contact resistances at the near-end and far-end nodes of nano interconnect nets can cause complications in boundary conditions. To resolve this issue, a model for analysis of crosstalk noise in CNT interconnects using FDTD methodology was published (Liang et al., 2012; Pu et al., 2009).

As shown in Figure 4.14, the two coupled CNT-based interconnect nets are termed aggressor and victim net having a per unit length scattering resistances r_{sagg}, r_{svic}; kinetic inductances l_{kagg}, l_{kvic}; magnetic inductances l_{eagg}, l_{evic}; quantum capacitances c_{qagg}, c_{qvic}; electrostatic capacitances c_{eagg}, c_{evic}; and load capacitances C_{Lagg}, C_{Lvic}, respectively. The position along the aggressor and victim interconnect net and time is denoted z *and* t (Agrawal et al., 2018; Kumar et al., 2016).

From the telegrapher's equations, the expression for coupled MWCNT transmission line becomes

$$\frac{\mathrm{d}}{\mathrm{dz}} \boldsymbol{V}(z, t) + \boldsymbol{RI}(z, t) + \boldsymbol{L}\frac{\mathrm{d}}{\mathrm{dt}}\boldsymbol{I}(z, t) = 0 \tag{4.25}$$

$$\frac{\mathrm{d}}{\mathrm{dz}} \boldsymbol{I}(z, t) + \frac{\mathrm{d}}{\mathrm{dt}} \boldsymbol{CV}(z, t) = 0 \tag{4.26}$$

where $\boldsymbol{V}$ and $\boldsymbol{I}$ are a 2 x 1 column matrix defined as

$$\boldsymbol{V} = \begin{bmatrix} \mathrm{V}_1 \\ \mathrm{V}_2 \end{bmatrix}, \boldsymbol{I} = \begin{bmatrix} \mathrm{I}_1 \\ \mathrm{I}_2 \end{bmatrix}, \quad R = \begin{bmatrix} r_{sagg} & 0 \\ 0 & r_{svic} \end{bmatrix}, \quad \boldsymbol{L} = \begin{bmatrix} l_{kagg} + l_{eagg} & l_{12} \\ l_{12} & l_{kvic} + l_{evic} \end{bmatrix} \text{ and}$$

$$\boldsymbol{C} = \begin{bmatrix} \left(\frac{1}{c_{qgg}} + \frac{1}{c_{eagg}}\right)^{-1} + c_{12} & -\,c_{aggvic} \\ -\,c_{aggvic} & \left(\frac{1}{c_{qvic}} + \frac{1}{c_{evic}}\right)^{-1} + c_{aggvic} \end{bmatrix}$$

The interconnect net of length l is driven by a resistive driver at $z = 0$ and terminated by a capacitive load at $z = l$. The net is discretized into Nz uniform segments of length $\Delta z = \frac{l}{Nz}$. The voltage and current solution points are discretized along the net as shown in Figure 4.15 (Figure 4.15).

With the application of finite central difference approximation; the equations obtained are

$$\frac{\boldsymbol{V}_{k+1}^{n+1} - \boldsymbol{V}_k^{n+1}}{\Delta z} + \boldsymbol{L}\frac{\boldsymbol{I}_k^{n+3/2} - \boldsymbol{I}_k^{n+1/2}}{\Delta t} + \boldsymbol{R}\frac{\boldsymbol{I}_k^{n+3/2} + \boldsymbol{I}_k^{n+1/2}}{2} = 0 \tag{4.27}$$

$$\boldsymbol{I}_k^{n+3/2} = \boldsymbol{EFI}_k^{n+1/2} + \boldsymbol{E}(\boldsymbol{V}_k^{n+1} - \boldsymbol{V}_{k+1}^{n+1}) \; for \;\; k = 1, 2, \ldots. N_z \tag{4.28}$$

where $= \left[\frac{\Delta z}{\Delta t}\boldsymbol{L} + \frac{\Delta z}{2}\boldsymbol{R}\right]^{-1}, \boldsymbol{F} = \left[\frac{\Delta z}{\Delta t}\boldsymbol{L} - \frac{\Delta z}{2}\boldsymbol{R}\right]$

$$\frac{\boldsymbol{I}_k^{n+1/2} - \boldsymbol{I}_{k-1}^{n+1/2}}{\Delta z} + \boldsymbol{C}\frac{\boldsymbol{V}_k^{n+1} - \boldsymbol{V}_k^n}{\Delta t} = 0 \tag{4.29}$$

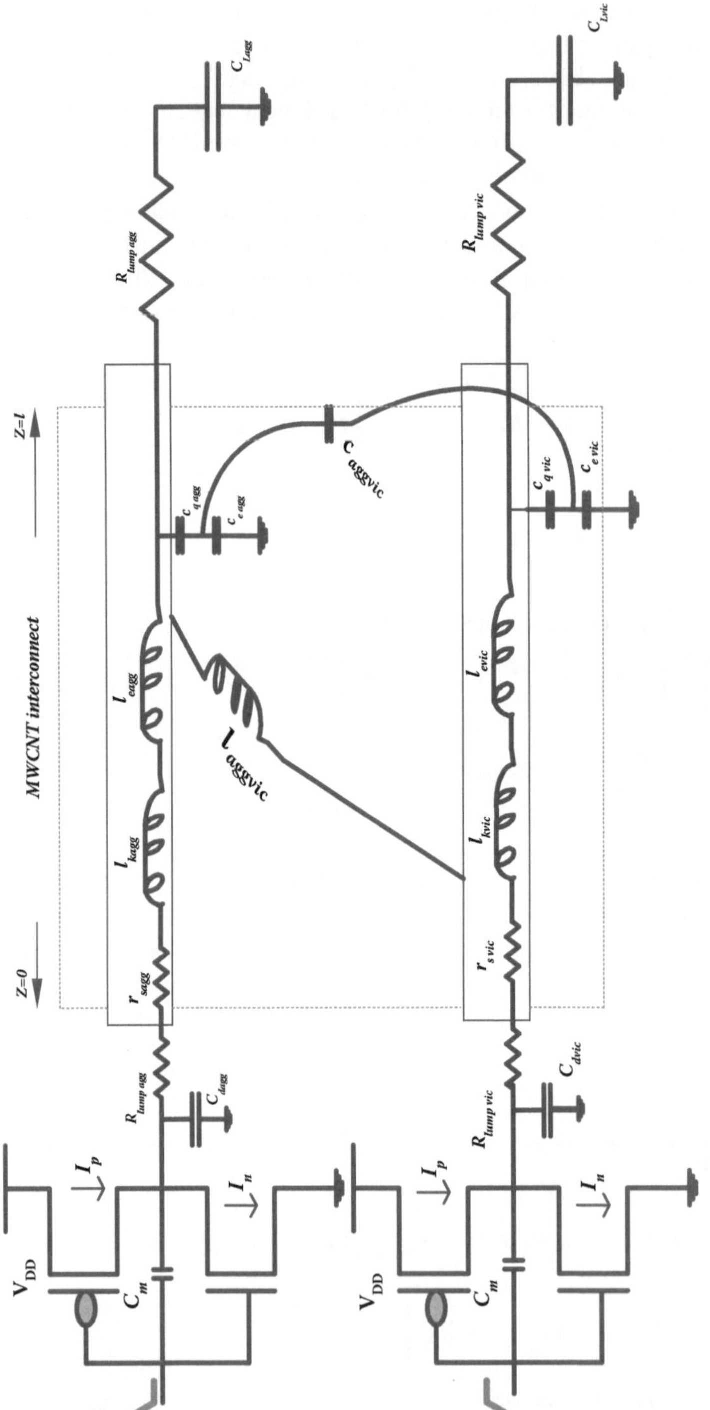

FIGURE 4.14 Coupled MWCNT aggressor (Line 1) victim (Line 2) interconnect nets driven by FET gate.

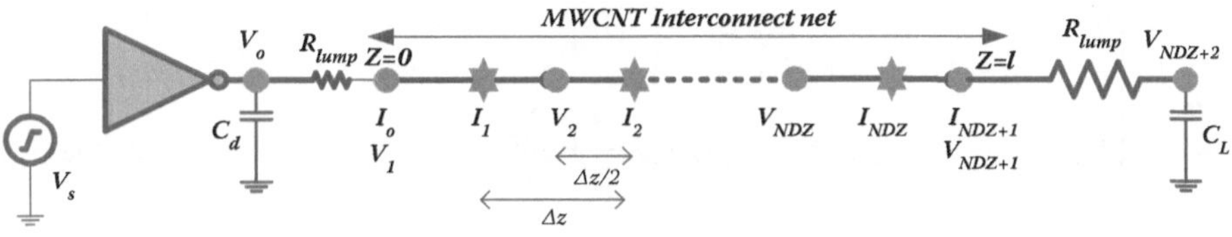

FIGURE 4.15 Space discretization of FDTD technique on MWCNT interconnects net

$$V_k^{n+1} = V_k^n + D(I_{k-1}^{n+1/2} - I_k^{n+1/2})\, for\ k = 2, 3, \ldots N_z. \tag{4.30}$$

where $D = \left[\frac{\Delta z}{\Delta t} C\right]^{-1}$

Assimilation of Near-End Boundary Conditions

At the near-end boundary conditions, putting $k = 1$ into the expression becomes

$$V_1^{n+1} = V_1^n + 2D(I_0^{n+1/2} - I_1^{n+1/2}). \tag{4.31}$$

The source current I_0 at the $(n + (1/2))$ time interval is computed by taking the average of values at n and $n + 1$, the expression becomes

$$V_1^{n+1} = V_1^n + 2D\left(\frac{I_0^{n+1} + I_0^n}{2} - I_1^{n+1/2}\right). \tag{4.32}$$

The I_0 denotes the driving current of FET. The expression for driving current is obtained by applying the KCL at the near-end boundary condition

$$V_0^{n+1} = V_0^n + A\left[\frac{C_m}{\Delta t}(V_s^{n+1} - V_s^n) + I_p^{n+1} - I_n^{n+1} - I_0^n\right] \tag{4.33}$$

$$V_1^{n+1} = BV_1^n + 2BD\left[\frac{V_0^{n+1}}{2R_{lump}} + \frac{I_0^n}{2} - I_1^{n+1/2}\right] \tag{4.34}$$

$$I_0^{n+1} = \frac{1}{R_{lump}}[V_0^{n+1} - V_0^{n+1}] \tag{4.35}$$

$A = \left[\frac{C_m + C_d}{\Delta t}\right]^{-1}$, $B = \left[U + \frac{D}{R_{lump}}\right]^{-1}$

C_m is the drain-to-gate coupling capacitance; C_d is the diffusion capacitance of FET.

The current flowing in pFET is denoted by I_p and in nFET is denoted by I_n. Applying the nth power law model, these currents can be expressed as

$$I_p = \begin{cases} 0; & V_s \geq V_{DD} - |V_{Tp}| \ (cutoff) \\ K_{lp}(V_{DD} - V_s - |V_{tp}|)^{\alpha_p/2}(V_{DD} - V_0); & V_0 > V_{DD} - V_{DSATp} \ (linear) \\ K_{sp}(V_{DD} - V_s - |V_{tp}|)^{\alpha_p}\left(1 + \sigma_p(V_{DD} - V_0)\right); & V_0 \leq V_{DD} - V_{DSATp} \ (saturation) \end{cases}$$

$$I_n = \begin{cases} 0; & V_s \leq |V_{tn}| \ (cutoff) \\ K_{ln}(V_s - |V_{tn}|)^{\alpha_n/2}(V_0); & V_0 < V_{DSATn} \ (linear) \\ K_{sn}(V_s - V_{tn})^{\alpha_n}(1 + \sigma_n(V_0)); & V_0 \geq V_{DSATn} \ (saturation) \end{cases}$$

The linear region transconductance parameter is denoted by K_{ln} and K_{lp}, saturation region transconductance parameter is K_{sn} and K_{sp}, threshold voltage is V_{tn} and V_{tp}, velocity saturation index is α_n and α_p and the drain conductance parameter is σ_n and σ_p for NFET and PFET, respectively.

Far-End Boundary Condition: At the far-end boundary conditions, put $k = Nz + 1$ and $Nz + 2$ and then the expression for voltage at the receiver end can be computed.

For $k = Nz + 1$, the voltage expression becomes

$$\boldsymbol{V}_{Nz+1}^{n+1} = \boldsymbol{V}_{Nz+1}^{n} + 2\boldsymbol{D}\left(\boldsymbol{I}_{Nz}^{n+1/2} - \frac{\boldsymbol{I}_{Nz+1}^{n+1} + \boldsymbol{I}_{Nz+1}^{n}}{2}\right). \tag{4.36}$$

To calculate the value of the output current I_{Nz+1}, apply KCL

$$V_{Nz+1} - V_{Nz+2} = R_{lump} I_{Nz+1} \tag{4.37}$$

On discretizing the value we get

$$\boldsymbol{I}_{Nz+1}^{n+1} = \frac{1}{\boldsymbol{R}_{lump}}[\boldsymbol{V}_{Nz+1}^{n+1} - \boldsymbol{V}_{Nz+2}^{n+1}] \tag{4.38}$$

Thus, the far-end voltage V_{Nz+1} and the load voltage V_{Nz+2} are given as

$$\boldsymbol{V}_{Nz+1}^{n+1} = \boldsymbol{B}\boldsymbol{V}_{Nz+1}^{n} + 2\boldsymbol{B}\boldsymbol{D}\left(\frac{\boldsymbol{V}_{Nz+2}^{n+1}}{2\boldsymbol{R}_{lump}} + \boldsymbol{I}_{Nz}^{n+1/2} - \frac{\boldsymbol{I}_{Nz+1}^{n}}{2}\right). \tag{4.39}$$

$$\boldsymbol{V}_{Nz+2}^{n+1} = \boldsymbol{V}_{Nz+2}^{n} + \frac{\Delta t}{C_L}\boldsymbol{I}_{Nz+1}^{n}. \tag{4.40}$$

4.4.3 FDTD Model for Crosstalk Analysis of Graphene Nanoribbon (GNR) Interconnects

Qian et al. (2016) found that while analyzing the crosstalk effects of coupled MLGNR, they considered the mean free path parameter independent of width by taking perfectly smooth edges of MLGNR. But practically all GNRs exhibit edge roughness (Areshkin et al., 2007) due to which electron scattering increases, hence decreasing the overall MFP and rise in resistivity. This section discusses a realistic model for analyzing MLGNR interconnects performance based on the FDTD method. In addition to this, a nonlinear FET driver is taken to drive the MLGNR aggressor and victim net.

Consider a test bench architecture of the MLGNR-based interconnect net of l length. At the near end, the FET gate is used as a driver and at the far end a capacitance is used to terminate the load. Discretizing the total interconnect net length into Nz uniform sections of space step Δz and complete simulation time into n uniform segments of time step Δt. To find the value of n, decimate the complete simulation time by Δt. While the value of Δt; the time step is found out by the Courant stability condition. For stable operation, the allowed maximum time step is $\Delta t_{max} = \frac{\Delta z}{v_{max}}$, where v_{max} is the maximum phase velocity. As shown in the figure, the voltage and current are discretized along the interconnect length (Figure 4.16).

Applying the telegrapher's equations, the expression for coupled MLGNR transmission lines becomes

$$\frac{d}{dz}\boldsymbol{V}(z, t) + \boldsymbol{RI}(z, t) + \boldsymbol{L}\frac{d}{dt}\boldsymbol{I}(z, t) = 0 \tag{4.41}$$

$$\frac{d}{dz}\boldsymbol{I}(z, t) + \frac{d}{dt}\boldsymbol{CV}(z, t) = 0 \tag{4.42}$$

where $\boldsymbol{V}$ and $\boldsymbol{I}$ are a $NX1$ column matrix and line parasitic is expressed in the NXN form.

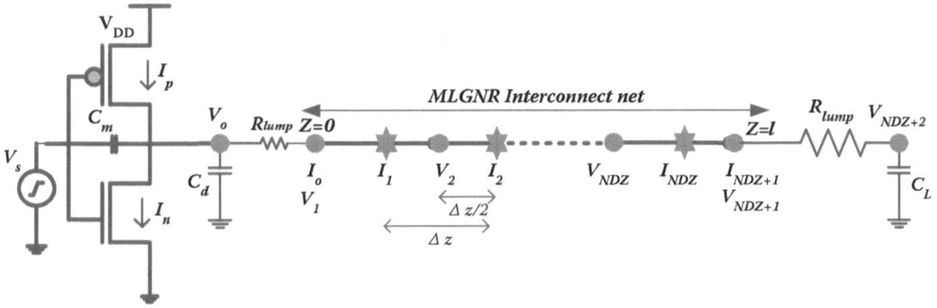

FIGURE 4.16 Space discretization of FDTD technique on MLGNR interconnects net.

The voltage $\boldsymbol{V}$ is given as $[V_1 V_2 \ldots \ldots V_{N-1} V_N]^T$; resistance $\boldsymbol{R}$ is $\begin{bmatrix} r_{s1} & 0 & 0 & \cdots & \cdots \\ 0 & r_{s2} & 0 & \cdots & \cdots \\ 0 & 0 & r_{s3} & \cdots & \cdots \\ \vdots & \vdots & \vdots & \cdots & \cdots \\ \vdots & \vdots & \vdots & 0 & r_{sN} \end{bmatrix}$; with r_{sN} as scattering resistance inductance $\boldsymbol{L}$ matix is $\begin{bmatrix} l_{k1}+l_{e1} & l_{12} & l_{13} & .. & .. \\ l_{21} & l_{k2}+l_{e2} & l_{23} & . & . \\ l_{31} & l_{32} & l_{k3}+l_{e3} & . & . \\ . & . & . & . & . \\ . & . & . & l_{N-1,N} & l_{kN}+l_{eN} \end{bmatrix}$; with $l_{N-1,N}$ as mutual inductance between the lines N-1 and N.

By applying a finite difference approximation, the line voltages and currents can be determined by applying the KCL to the boundary terminal voltages and currents can be computed as

$$V_0^{n+1} = V_0^n + \boldsymbol{A}\left[\frac{C_m}{\Delta t}\left(V_s^{n+1} - V_s^n\right) + I_p^{n+1} - I_n^{n+1} - I_0^n\right] \tag{4.43}$$

$$V_1^{n+1} = \boldsymbol{B}\boldsymbol{V}_1^n + 2\boldsymbol{B}\boldsymbol{D}\left[\frac{V_0^{n+1}}{2\boldsymbol{R}_{lump}} + \frac{\boldsymbol{I}_0^n}{2} - \boldsymbol{I}_1^{n+1/2}\right] \tag{4.44}$$

$$V_k^{n+1} = \boldsymbol{V}_k^n + \boldsymbol{D}\left[\,\boldsymbol{I}_{k-1}^{n+1/2} - \boldsymbol{I}_k^{n+1/2}\right] \text{ for} k = 2, 3, \ldots Nz \tag{4.45}$$

$$V_{Nz+2}^{n+1} = V_{Nz+2}^n + \frac{\Delta t}{C_L}\boldsymbol{I}_{Nz+1}^n \tag{4.46}$$

$$V_{Nz+1}^{n+1} = \boldsymbol{B}\boldsymbol{V}_{Nz+1}^n + 2\boldsymbol{B}\boldsymbol{D}\left[\frac{\boldsymbol{V}_{Nz+2}^{n+1}}{2R_{lump}} + \boldsymbol{I}_{Nz}^{n+1/2} - \frac{\boldsymbol{I}_{NZ+!}^n}{2}\right] \tag{4.47}$$

$$\boldsymbol{I}_0^{n+1} = \frac{1}{\boldsymbol{R}_{lump}}\left[\boldsymbol{V}_0^{n+1} - \boldsymbol{V}_1^{n+1}\right] \tag{4.48}$$

$$\boldsymbol{I}_k^{n+3/2} = \boldsymbol{E}\boldsymbol{F}\boldsymbol{I}_k^{n+1/2} + \boldsymbol{E}\left[\boldsymbol{V}_k^{n+1} - \boldsymbol{V}_{k+1}^{n+1}\right] for\ k = 1, 2, \ldots Nz \tag{4.49}$$

$$\boldsymbol{I}_{Nz+1}^{n+1} = \frac{1}{\boldsymbol{R}_{lump}}\left[\boldsymbol{V}_{Nz+1}^{n+1} - \boldsymbol{V}_{Nz+2}^{n+1}\right] \tag{4.50}$$

where $= \left[\frac{C_m + C_d}{\Delta t}\right]^{-1}$, $\boldsymbol{B} = \left[\boldsymbol{U} + \frac{\boldsymbol{D}}{\boldsymbol{R}_{lump}}\right]^{-1}$, $\boldsymbol{D} = \left[\frac{\Delta z}{\Delta t}\boldsymbol{C}\right]^{-1}$, $\boldsymbol{E} = \left[\frac{\Delta z}{\Delta t}\boldsymbol{L} + \frac{\Delta z}{2}\boldsymbol{R}\right]^{-1}$,

$F = \left[\frac{\Delta z}{\Delta t} L - \frac{\Delta z}{2} R\right]$, C_m is a drain to the gate coupling capacitance and C_d is the drain diffusion capacitance.

4.5 CROSSTALK RESULT ANALYSIS AND DISCUSSIONS

For deep sub-micron technology, the impact of crosstalk is a crucial parameter of determination. In view of the shrinking interconnect width with progress in process technology at the end of road map, it is apparent that crosstalk propagation delay becomes significant. In the case of on-chip nano interconnects, both crosstalk-induced noise and crosstalk-induced delay are gradually becoming paramount consideration factors. In this section, a detailed comparative analysis of crosstalk-induced delay as well as noise for different types of interconnect materials is provided. The crosstalks investigations are carried out for different interconnect lengths and different temperatures.

With the intention of conveniently investigating the impact of crosstalk on coupled dual line identical interconnecting wire, a test bench architecture comprising a driver interconnect load (DIL) structure, shown in Figure 4.4, is considered for analysis. For the simulation, a CMOS inverter logic gate is taken to drive the coupled aggressor and victim interconnect lines. From previous research, it has been observed that delay under the influence of crosstalk is increased by 80% if we assume driver buffer to be resistive instead of CMOSFET. Hence, a resistive driver may result in erroneous estimation of crosstalk delay. To analyze and investigate the crosstalk-induced delay, internal coupling capacitance (C_{CM}) is considered in between the aggressor and victim net. For simulation purposes, here we consider the CMOS driver operating at a clock speed of 0.1 GHz, with a supply of V_{dd} = 0.8V and terminated with a load capacitance C_L of 0.20fF.

As manifested from the discussions in Sections 4.2 and 4.3, in an electrostatically and magnetically coupled nanointerconnect structure, the signal transition on an aggressor net leaves an impact on the voltage and signal switching of the victim net. This phenomenon event is called crosstalk. Aggressor net is the one that is responsible for the occurrence of this event, while the wire net that is affected is called the victim net. The crosstalk is mainly categorized into two types: dynamic crosstalk and functional crosstalk. Depending on the switching of inputs at that aggressor driver buffer and also on the potential applied and switching of inputs at the victim driver buffer, there can be different scenarios, as illustrated in Table 4.1. For the analyses of dynamic crosstalk, both aggressor and victim wire nets are made to switch signals, whereas for the analyses of functional crosstalk the switching of signals is made only on an aggressor wire net while the victim net is held static at either a high (hi) logic level or low (lo) logic level. Due to coupling between the nets, an undesirable noise in the form of a voltage glitch is observed at the victim affected node. This glitch might cause a logic error on the transmitted interconnect line if the amplitude of the voltage glitch crosses the threshold limit. Moreover, signal line overshoot/undershoot is the result of the crosstalk effect in any interconnect system.

TABLE 4.1
Crosstalk scenarios under different switching conditions of aggressor and victim net

Crosstalk Condition	Aggressor Net Logic Switching	Victim Net Logic Switching	Cases	Affect
Functional Crosstalk	Low to High	Held static at LOGIC 1	FC-I	Overshoot
	Low to High	Held static at LOGIC 0	FC-II	Rise-time glitch
	High to Low	Held static at LOGIC 1	FC-III	Fall-time glitch
	High to Low	Held static at LOGIC 0	FC-IV	Undershoot
Dynamic Crosstalk	Low to High	Low to High	DC-I	Decrease in rise time
	Low to High	High to Low	DC-II	Increase in fall time
	High to Low	Low to High	DC-III	Increase in rise time
	High to Low	High to Low	DC-IV	Decrease in fall time

TABLE 4.2
Average functional crosstalk-induced noise values for Cu, CNT and GNR interconnects under different aggressor and victim switching conditions

Switching Cases	Type of Interconnect (mV)			Result Outcome at Victim Node	Impact
	Cu	CNT	GNR		
FC-I	901.35	862.35	825.25	Overshoot	False prediction of logic, causing reliability issue
FC-II	110	71.48	27.43	Rise glitch	Functionality failure
FC-III	738.35	731.66	775.66	Fall glitch	Functionality failure
FC-IV	101.10	64.56	14.89	Undershoot	False prediction of logic, causing reliability issue

Table 4.2 manifests the average functional crosstalk-induced noise values under four different cases investigated for the three types of interconnects discussed in this chapter, namely Cu interconnects net, CNT interconnects net and GNR interconnects net. For each switching case, the value of the average crosstalk-induced noise glitch at the output of the victim net is taken for the simulation environment at

TABLE 4.3
Average dynamic crosstalk-induced delay values for Cu, CNT and GNR interconnects under different aggressor and victim switching conditions

Switching Cases	Delay for Interconnect (nsec)			Result Outcome at Victim Mode	Impact
	Cu	CNT	GNR		
DC-I	1.38	2.15	1.1	Decrease in rise time	Timing violation
DC-II	1.94	2.73	1.73	Increase in fall time	Timing violation
DC-III	2.15	2.72	1.74	Increase in rise time	Timing violation
DC-IV	1.38	2.12	1.11	Decrease in fall time	Timing violation

the ambient temperature, and the driver and load buffer is modeled using the HSpice model from the predictive technology model (PTM) source file at 32nm technology node (Cao, 2008). The neighboring nets taken for analysis are considered distributed RLC wire models.

Table 4.3 manifests the average dynamic crosstalk-induced noise values under four different cases for the three types of interconnects discussed in this chapter, namely Cu interconnects net, CNT interconnects net and GNR interconnects net.

It is evident from the results of Table 4.2 that GNR-based interconnects demonstrate the lowest glitch magnitude as compared to Cu and CNT interconnects. In the case of FC-I, the average percentage mitigation in the value of a functional crosstalk-induced overshoot glitch for GNR interconnects is 4.62% of CNT and 8.37% of Cu. For the case of FC-II, the average percentage mitigation in value of a functional crosstalk-induced overshoot glitch for GNR interconnects are 62.66% of CNT value and 74.44% with respect to the Cu interconnect net. Similarly, the average percentage reductions in value of a functional crosstalk-induced overshoot glitch for GNR interconnects with respect to CNT interconnects and Cu net for the case of FC-III is 5.42% and 0.13%, respectively. Meanwhile, for the case of FC-IV, the average percentage reductions in value of functional crosstalk-induced overshoot positive glitch for GNR interconnects is 63.34% and 72.81%, respectively. Rise and fall glitches have a prominent impact on the logical, functional behavior of the combinational and sequential circuit, while the circuit's reliability is affected and is at a potential risk due to overshoot and undershoot obtained.

The result obtained in Table 4.3 demonstrates that average percentage reduction in the dynamic crosstalk-induced delay for GNR interconnects with respect to CNT and Cu interconnects under different cases obtained from the aggressor and victim switching combination. A transient analysis is done to generate these simulated results. For the case of DC-I, the mitigation in average percentage value of dynamic crosstalk-induced delay for GNR with regard to CNT interconnects and Cu interconnects is 96.33% and 24.24%, respectively, whereas for the case of DC-II the average reduction in the dynamic crosstalk-induced delay for GNR with regard to CNT interconnects and Cu interconnects is 55.33% and 10.94%, respectively. Similarly, the average reduction in the dynamic crosstalk-induced delay for GNR with regard to CNT interconnects and Cu interconnects is 55.78% and 22.45%, respectively, for the case of DC-III. Moreover, for the case DC-IV, the average percentage reduction in the dynamic crosstalk-induced delay for GNR interconnects with regards to CNT and Cu interconnects is obtained as 91.87% and 22.34%, respectively. To the best of our knowledge, from different literature reviews, it is manifested that crosstalk-induced delay varies with temperature fluctuations and also changes with variation in length. With a rise in temperature, the crosstalk-induced delay increases, because of a reduction in the effective mean free path of interconnect wire.

With regards to functional crosstalk, for a driver interconnect load structure, on the basis of the height and duration of the spurious signal, the next digital repeater gate might interpret it as an actual valid signal, thereby causing a logic error that can be further propagated. This error can be enduring if it affects the circuit's sequential block. In combinational blocks, only transient errors will appear. The main waveform parameters influencing the effect of spurious signals through logic are its amplitude and its width.

Tables 4.4 and 4.5 illustrate the overshoot/undershoot width and peak voltage variation of Cu, CNT and GNR interconnects for change in wire length over a temperature range of 233 K to 450 K. The width of the overshoot/undershoot is computed by calculating the overshoot/undershoot area under the waveform by sampling the waveform into small rectangular sections. From the tabulated data, it is inferred that the rise in peak overshoot voltage takes place with an increase in coupling capacitance value in between the on-chip nano interconnects. The value decreases with a rise in conductance and capacitance to ground value for interconnects. The GNR interconnects exhibits superior performance compared to the other two counterparts due to less coupling capacitance and high conductance and more value of capacitance to ground. The consequence of this is a lower peak overshoot voltage in GNR-based interconnects than in Cu and CNT.

The results from Table 4.5 show the obtained overshoot/undershoot peak values for different interconnect lengths at various temperatures. The peak values are obtained for Cu, CNT and GNR interconnects.

The tabulated data obtained in Table 4.5 show the peak values of overshoot/undershoot for varying wire lengths at different temperatures. The table also points out that for the case of GNR interconnects, the peak overshoot voltage value is below 16% of the supply voltage V_{DD} and also it is observed that there is no considerable rise in its value with the variation in temperature and length of the interconnecting wire. However, on the other side, if we consider the case of CNT interconnects, then the peak overshoot voltage value increases by 24%. Furthermore, on analyzing the tabulated data obtained for Cu interconnects, the peak overshoot voltage value reaches almost 45% of the power supply voltage. Pondering over the trend followed, it can be inferred that because of higher conductance the performance of carbon-based interconnect wire structures improves and in some situations it may reache

TABLE 4.4
Crosstalks induced overshoot/undershoot width of Cu, CNT and GNR interconnect for different length over temperature range of 233 K to 450 K

Overshoot Width (ps) Length (μm)	Cu	CNT	GNR
Temperature = 233 K			
10	26.95	17.86	23.36
20	58.67	32.37	46.44
30	102.34	47.07	73.64
40	157.76	61.89	104.92
50	221.54	76.93	140.29
Temperature = 300 K			
10	30.47	17.93	24.10
20	76.25	32.76	49.51
30	140.53	47.87	80.57
40	223.36	63.44	117.23
50	324.77	105.69	158.40
Temperature = 378 K			
10	39.44	18.14	24.99
20	109.06	33.58	53.06
30	211.90	49.60	88.78
40	347.94	66.39	131.66
50	517.12	83.95	181.94
Temperature = 450 K			
10	50.44	18.42	25.88
20	149.17	34.66	56.50
30	299.05	51.93	96.44
40	498.87	70.38	145.19
50	752.53	90.14	202.92

even lower than 20%. This is contrary to the trend observed in conventional Cu-based interconnect wire structures, where the value of peak overshoot voltage rises in a linear fashion with a variation in interconnect length as well as temperature. For some cases of graphene-based nano interconnects, particularly top-contacted nanoribbons, there is a dip in conductance as very few top layers participate in the current conduction, but still this drop is not as significant as the Cu interconnect. Apart from all this, it is also noticed from the data of Tables 4.4 and 4.5 that the width of the overshoot is greater in the case of GNR than CNT-based interconnects, whereas the peak of CNT is greater than GNR-based interconnects. The Cu interconnects exhibit larger widths and higher peaks due to the capacitance to ground value. Hence, the average failure rate is greater for Cu interconnects.

TABLE 4.5
Crosstalk induced overshoot/undershoot peak voltage of Cu, CNT and GNR interconnect for different length over temperature range of 233 K to 450 K

Overshoot Peak (volt) Length (μm)	Cu	CNT	GNR
Temperature = 233 K			
10	0.933	0.839	0.788
20	0.945	0.858	0.798
30	0.967	0.867	0.803
40	0.987	0.871	0.804
50	0.991	0.875	0.805
Temperature = 300 K			
10	0.934	0.843	0.789
20	0.967	0.862	0.799
30	0.981	0.870	0.803
40	0.992	0.877	0.805
50	1.003	0.879	0.807
Temperature = 378 K			
10	0.938	0.848	0.789
20	0.969	0.867	0.799
30	0.988	0.874	0.804
40	0.999	0.879	0.807
50	1.008	0.880	0.809
Temperature = 450 K			
10	0.940	0.851	0.790
20	0.978	0.868	0.801
30	0.991	0.874	0.806
40	1.004	0.880	0.810
50	1.013	0.884	0.811

In the subsequent paragraphs, by using capacitively coupled interconnect wire lines, the in-phase and out-phase delays are observed at victim nets with respect to the switching in aggressor nets. It will be proved from the results that propagation delay under the influence of crosstalk increases for global interconnects level. The result obtained will prove that in-phase delay is less compared to the out-phase delay for any specific interconnect length considered. The primary reason behind this is the effect of the Miller capacitance. The presence of the Miller capacitive effect results in nearly doubling of the internal coupling capacitance (C_{CM}). It is noticed that under the influence of crosstalk out-phase transitions, the Miller coupling factor tends to be a value of 2.

For a very high frequency of operation, the tightly packed interconnects generates transient crosstalk. This crosstalk, as mentioned earlier, has a strong impact on signal propagation delay and in some cases might result in logic or functional failure. Therefore a model for analytically analyzing the crosstalk effects occurring in nano interconnects was discussed in the section for Cu, CNT and GNR interconnects.

Past history reflects that numerous analytical models have been recommended by researchers (Khursheed et al. 2020). The model discussed in this chapter is formulated by finite difference time domain (FDTD) methodology for tDIL system. The voltage and current values can be predicted at any specific point of interconnect net using FDTD methodology. Here, the FET driver is modeled by the nth power law by taking into consideration finite drain conductance parameter. Furthermore, for Cu interconnecting wires the short channel effects including velocity saturation and finite drain conductance parameter are also considered in the proposed model. To test the sturdiness of the FDTD model, the computational result obtained using this model is validated by comparing it with results obtained from HSPICE simulations. The percentage error is also computed while measuring the crosstalk-induced peak voltage and its timing instances with regards to HSPICE simulation results. For analysis purposes, Figure 4.11 is used as a test bench. Here, two coupled interconnect lines (aggressor and victim) are driven by FET gates. The aggressor and victim nets are modeled using a distributed RLC parasitic. The transient response obtained for the DIL structure is illustrated in Figure 4.11, and are compared using HSPICE. For the symmetric driving capability of the driver, the width of the PFET is chosen as twice the NFET width. The time domain response is performed at a 32 nm technology node at the global-level interconnect length of 1 mm. The input transition time and supply voltage are considered as 10 ps and 0.9 V, respectively. The dimensions of the interconnect line are considered by two assumptions: (1) the space between the two interconnects is equal to the width of interconnecting; and (2) the height from the ground plane is equal to the thickness of the line. The resistivity of the copper material and the relative permittivity of the inter layer dielectric medium are chosen as 2.2 (μΩ-cm) and 2.2, respectively. The interconnect line width and the aspect ratio are considered as 0.22 μm and 3, respectively. The load capacitance, C_L, is considered as 2 fF. The transient response comparison of Cu, CNT and GNR on-chip interconnects using the CMOS driver is tabulated in Tables 4.6 to 4.11.

Table 4.6 shows the variation in crosstalk-induced noise voltage level with an increase in length of the coupled interconnect wires. The noise voltage value is computed using the FDTD model and compared with the HSPICE value.

Table 4.7 demonstrates the propagation delay of the Cu wire under the influence of dynamic crosstalk when the neighboring nets, aggressor and victim, switch simultaneously. A variation in signal propagation delay is observed under the influence of dynamic crosstalk if adjoining aggressor switches are either in the same direction (in-phase) or in the opposite direction (out-phase) to the signal flow direction of victim.

In Table 4.8, the observations of crosstalk delay variations in the case of carbon nanotube (CNT) based interconnects are made for different interconnect lengths. The length is varied from 200 μm to 1,200 μm, with a fixed spacing of 2 nm for this case. The in-phase delay as well as out-phase delay is calculated with the FDTD

TABLE 4.6
Crosstalk noise voltage variations with interconnect length in Cu interconnects using HSPICE and the FDTD model

Interconnect Length (μm)	Crosstalk Noise (V)		
	HSPICE	FDTD	% Error
200	0.08	0.08	0.00
400	0.14	0.14	0.00
600	0.23	0.22	4.13
800	0.28	0.27	3.44
1,000	0.32	0.30	3.01
1,200	0.36	0.36	0.00

TABLE 4.7
Crosstalk delay variations with interconnect length in Cu interconnects using HSPICE and the FDTD model

Interconnect Length (μm)	In-Phase Delay (ps)			Out-Phase Delay (ps)		
	HSPICE	FDTD	% Error	HSPICE	FDTD	% Error
200	52.67	51.22	2.66	231.80	229.02	1.20
400	73.70	72.18	2.06	368.71	366.03	0.41
600	93.44	91.18	2.50	521.07	515.05	1.15
800	113.94	110.68	2.81	716.00	709.08	0.96
1,000	146.17	144.01	1.50	960.50	953.08	1.08
1,200	166.96	165.03	1.16	1263.40	1255.09	0.66

model mentioned in Section 4.4.2 and then the computed values are benchmarked with HSPICE simulation results.

Table 4.9 demonstrates the variation in crosstalk-induced noise voltage level with an increase in length of the coupled interconnect wires. The value of the peak noise voltage rises monotonically with an increase in interconnect length from local level to global level. In addition to this, as the crosstalk noise is also a function of coupling capacitance, hence the appropriate value is taken of coupling capacitance because it directly influences the noise factor.

Table 4.10 of the GNR net shows a similar observation trend as Table 4.8 for interconnect wire length ranging from 200 μm to 1,200 μm at 1 ns transition and 2 nm spacing between the adjoining aggressor and victim net. It is worthwhile to mention that a significant attenuation in value of crosstalk-induced delay is observed for carbon-based CNT/GNR interconnects compared to conventional Cu interconnects.

TABLE 4.8
Crosstalk delay variations with interconnect length in CNT interconnects using HSPICE and the FDTD model

Interconnect Length (μm)	In-Phase Delay (ps)			Out-Phase Delay (ps)		
	HSPICE	FDTD	% Error	HSPICE	FDTD	% Error
200	17.13	17.00	0.007	103.80	100.43	0.324
400	20.38	18.12	0.023	168.71	155.66	0.073
600	25.95	28.95	-0.103	273.07	272.07	0.003
800	25.97	25.97	0.00	416.00	409.99	0.014
1,000	27.87	27.09	0.027	656.50	655.50	0.0015
1,200	31.11	28.78	0.074	863.40	861.12	0.0026

TABLE 4.9
Crosstalk noise voltage variations with interconnect length in CNT interconnects using HSPICE and the FDTD model

Interconnect Length (μm)	Crosstalk Noise (mV)		
	HSPICE	FDTD	% Error
200	152.50	152.50	0.00
400	181.32	181.32	0.00
600	174.66	172.11	0.010
800	203.11	202.43	0.003
1,000	211.18	208.99	0.103
1,200	362.51	362.51	0.00

The delay and crosstalk noise of the proposed model are measured by MATLAB and then compared with the HSPICE simulations. To carry out the FDTD analysis of a transmission line model, a computational domain has to be established. The computational domain for the transmission line is modeled by resistance, inductance and capacitance (RLC) elements with relative parameters of V and I. The V and I values are determined at each point in space and time within the computational domain. The resistance (R), inductance (L) and capacitance (C) parasitic of the line must be specified for each cell in the computational domain. In the case of the coupled line system, the mutual inductance and the coupling capacitance should also be included within the computational domain. After establishing the computational domain, the boundary conditions have to be specified at the near-end and far-end boundaries.

The rise and fall transition of the input signal is assumed to be 10 ps. In the two coupled interconnect lines, line 2 is considered a victim line and, accordingly, all

TABLE 4.10
Crosstalk delay variations with interconnect length in GNR interconnects using HSPICE and the FDTD model

Interconnect Length (μm)	In-Phase Delay (ps)			Out-Phase Delay (ps)		
	HSPICE	FDTD	% Error	HSPICE	FDTD	% Error
200	18.36	18.12	0.24	123.80	120.43	0.3166
400	24.817	23.12	1.69	268.71	256.66	0.044
600	35.95	35.95	0.00	282.07	282.07	0.00
800	25.47	24,76	−0.29	516.00	509.99	0.13
1,000	37.87	36.99	0.88	756.50	755.50	0.0014
1,200	38.811	38.78	0.007	963.40	961.12	0.0024

TABLE 4.11
Crosstalk noise voltage variations with interconnect length in GNR interconnects using HSPICE and the FDTD model

Interconnect Length (μm)	Crosstalk Noise (mV)		
	HSPICE	FDTD	% Error
200	142.35	135.50	0.048
400	158.32	158.32	0.00
600	164.66	162.11	0.015
800	199.11	199.43	−0.0016
1,000	201.18	201.99	0.00
1,200	262.51	262.51	0.00

the performance parameters such as propagation delay, noise peak voltage and its timing instances are evaluated on victim line 2. The functional crosstalk analysis is studied by switching the aggressor line 1 and keeping the victim line 2 in a quiescent mode. Later on, dynamic crosstalk analysis is studied by switching both aggressor and victim lines either in in-phase or out-phase.

The crosstalk-induced propagation delay is mainly determined by the value of the coupling capacitance (C_{cm}), which exists between aggressor and victim nets calculated by Eq. 4.1. The coupling capacitance demonstrates a linear proportionality with the length of interconnect wire taken and inverse proportionality with the spacing between adjacent lines. The accuracy of the proposed FDTD-based model is validated by SPICE simulations. With an increase in value of coupling capacitance, there is a significant increase in crosstalk peak noise voltage. The fluctuation in the case of out-phase delay with change in coupling capacitance is more than the fluctuation in the case of in-phase delay because of the Miller capacitance effect.

Finally, in comparing the results of Tables 4.6 and 4.9, it can be stated that the peak value of crosstalk-induced noise voltage for CNT-based interconnect wire is lower in magnitude than Cu. This is primarily due because the coupling capacitance in the case of MWCNT is smaller in value than the coupling capacitance in the case of the Cu counterpart. Furthermore, it is observed that generally for all three types of interconnects in increasing the length of wire, there is a gradual increase in noise width. The noise width is impacted primarily by the speed of the interconnecting wire; hence, it is manifested that the carbon-based CNT/GNR interconnect wire lines have a smaller noise width as juxtapose to conventional Cu wire.

For the case of functional crosstalk, the in-phase as well as out-phase dynamic switching conditions, the noise voltage peaks and crosstalk-induced delays are investigated for varying interconnect lengths ranging from 200 μm to 1,200 μm. From the obtained results an error percentage is calculated in the tables for crosstalk-induced noise peaks, in-phase delay and out-of-phase delay we interpret that the FDTD-based model results are in close agreement with the SPICE results. Hence, it is manifested from the error percentage values that the FDTD model holds well for a larger range of interconnect wire lengths. In addition to this, it is worthwhile to note that the average percentage error observed is less that 1% for in-phase delays and out-phase delays, whereas while computing the peak crosstalk noise voltage it is less than 2%. Hence, once again, these vivid facts show the results obtained for noise peak and crosstalk delay using the FDTD model are as precise and accurate as that retrieved from HSPICE simulations.

4.6 SUMMARY

- The prime intention of this chapter is to discuss the signal integrity challenges encountered when modeling state-of-he-art nano interconnects.
- As the feature size decreases due to the scaling of device dimensions, the reliability is compromised. The decrease in reliability is due to electromigration-induced problems.
- At a very high frequency of operation, the tightly packed nano interconnects produce transient crosstalk. Crosstalk-induced noise in the case of coupled interconnect lines is generally categorized into two types, namely *functional crosstalk* and *dynamic crosstalk.*
- In the case of functional crosstalk, whenever there is signal transition on the aggressor interconnect line, then the quiescent victim interconnect lines will experience a voltage spike, whereas in the case of dynamic crosstalk, both the aggressor net and victim net switch simultaneously either in the same direction or the opposite direction.
- Furthermore, analysis on how the crosstalk noise strongly influences the signal propagation delay and causes the circuit malfunction or functional failure is discussed.
- The crosstalk noise causes signal overshoot, undershoot and ringing effects; hence, it is enviable to develop a perfect pedagogical model for analytically analyzing the crosstalk effects occurring in nano interconnects.

- For this reason, the latter part of this chapter presents a precise and time-efficient model of FET-gate-driven coupled (Copper/CNT/GNR) nano interconnects in order to commensurate the crosstalk-incited performance analysis of nano interconnects.
- The model discussed is formulated by a finite difference time domain (FDTD) methodology for the DIL system by assimilating boundary conditions for all three types of interconnect materials. The FET driver is modeled by the nth power law model by considering the finite drain conductance parameter.
- By adopting the FDTD model, it is easy to specify the computational domain, the values for voltage can be conveniently measured at any point on the interconnect wire line and simulations can be carried out for a wide range of frequencies. The model discussed in this chapter uses dual-line coupled interconnects, but it can be easily extended for multiple interconnects.
- In the end, a detailed comparison of propagation delay and noise peak under the influence of crosstalk for Cu, CNT and GNR interconnects is done.
- Under the influence of functional crosstalk, in-phase and out-phase dynamic switching conditions, noise peak voltages and crosstalk-induced delays were examined at different interconnect wire lengths ranging from 200 to 1,200 μm. The results obtained using the FDTD model is in accordance to those of HSPICE simulations. The percentage error calculated between these readings is less than 2% for peak crosstalk noise and it is less than 1% for in-phase and out-phase dynamic crosstalk delays.

EXERCISES

Multiple-Choice Questions

Q1 The stability of the FDTD is limited by
  a Time step size
  b Space step size
  c Both a and b
  d None of the these.

Q2 In the FDTD model, the boundary conditions are derived in
  a An explicit manner
  b An implicit manner
  c Time domain
  d None of these.

Q3 The phase velocity of the signal in the transmission line depends on
  a Transmission line parasitic
  b Input switching
  c Coupling parasitic
  d All of the above.

Q4 In the FDTD model, the voltage and adjacent current solution point is separated by
  a Δz

b $\Delta z/2$
c $\Delta z/3$
d $\Delta z/4$

Q5 In dynamic out-phase crosstalk, the delay increases due to
a High resistance
b High inductance
c High capacitance
d All of the above.

Q6 In dynamic in-phase crosstalk, the delay decreases due to
a Low resistance
b Low inductance
c Low capacitance
d All of the above.

Q7 The accuracy of the FDTD model depends on
a The space step size
b The time step size
c The number of iterations
d All of the above.

Q8 The magnitude of the glitch caused depends upon the
a Slew of the aggressor net
b Victim net grounded capacitance
c Victim nets driver strength
d All of the above.

Q9 Overshoot glitch occurs when
a A rising aggressor couples to a victim net, which is steady low
b A falling aggressor couples to a victim net, which is steady low
c A rising aggressor couples to a victim net, which is steady high
d None of the these.

Q10 Which check is used for glitch magnitude and refers to DC noise limits on the input of a cell while ensuring proper logic functionality?
a DC noise margin
b AC noise margin
c Both a and b.
d None of the these.

Q11 An aggressor net is __________.
a A net that affects other nets
b A net that is affected by other nets
c A power net
d A ground net

Q12 Crosstalk affects __________.
a Timing
b Functionality

c Reliability
d All of these.

Q13 Crosstalk delay arises due to __________.
a Simultaneous switching of aggressor and victim nets
b Switching of only aggressor nets
c Switching of only victim nets
d None of these.

Short-Answer Questions

Q1 What is crosstalk?
Q2 How can you avoid crosstalk in long on chip interconnects?
Q3 How can a buffer be used as a repeater in the victim net to avoid crosstalk?
Q4 Differentiate between negative and positive crosstalk delay.
Q5 Which type of crosstalk is observed when the aggressor and victim lines switch simultaneously?
Q6 How does crosstalk affect setup and hold timing?
Q7 Discuss with the help of a schematic diagram how overshoot/undershoot analysis can be done.

Long-Answer Questions

Q1 Derive the boundary for the FDTD model of the Cu and CNT interconnects.
Q2 Explain different types of crosstalk mechanisms.
Q3 What is the difference between crosstalk noise and crosstalk delay?

ANSWERS TO MULTIPLE-CHOICE QUESTIONS

1. C	8. D
2. B	9. C
3. D	10. A
4. B	11. A
5. C	12. D
6. C	13. A
7. C	

REFERENCES

Agarwal, Kanak, Dennis Sylvester, and David Blaauw. "Modeling and Analysis of Crosstalk Noise in Coupled RLC Interconnects." *IEEE Transactions on Computer-Aided Design of Integrated Circuits and Systems* 25, no. 5 (2006): 892–901.

Agrawal, Yash and Rajeevan Chandel. "Crosstalk Analysis of Currentmodes Signaling Coupled RLC Interconnects Using FDTD technique." *IETE Technical Review* 33, no. 2 (2015): 148–159.

Agrawal, Yash, Mekala Girish, and Rajeevan Chandel. "A Unified Delay, Power and Crosstalk Model for Current Mode Signalling Multiwall Carbon Nanotube Interconnects." *Springer Circuits, Systems, and Signal Processing* 37, no. 4 (2018): 1359–1382.

Areshkin, Denis A., Daniel Gunlycke, and Carter T. White. "Ballistic Transport in Graphene Nanostrips in the Presence of Disorder: Importance of Edge Effects." *Nano Letters* 7, no. 1 (2007): 204–210.

Cao, Yu Kevin. Predictive Technology Model. (2008). [Online]. Available: http://ptm.asu.edu/.

Kaushik, Brajesh Kumar and Sankar Sarkar. "Crosstalk Analysis for a CMOS-gate-Driven Coupled Interconnects." *IEEE Transactions on Computer-Aided Design of Integrated Circuits and Systems* 27, no. 6 (2008): 1150–1154.

Khursheed, Afreen and Kavita Khare. "Optimized Buffer Insertion for Efficient Interconnects Designs." *International Journal of Numerical Modelling: Electronic Networks, Devices and Fields* 33, no. 5 (2020): e2748.

Kumar, Vobulapuram Ramesh, Brajesh Kumar Kaushik, and Amalendu Patnaik. "An Accurate FDTD Model for Crosstalk Analysis of CMOS-gate-Driven Coupled RLC Interconnects." *IEEE Transactions on Electromagnetic Compatibility* 56, no. 5 (2014): 1185–1193.

Kumar, Vobulapuram Ramesh, Brajesh Kumar Kaushik, and Amalendu Patnik. *Crosstalk in Modern on-Chip Interconnect: A FDTD Approach.* Singapore: Springer Brief in Applied Science and Technology, 2016.

Li, Xiao-Chun, Jun-Fa Ma, and Madhavan Swaminathan. "Transient Analysis of CMOS Gate Driven RLGC Interconnects Based on FDTD." *IEEE Transactions on Computer-Aided Design of Integrated Circuits and Systems* 30, no.4 (2011): 574–583.

Liang, Feng, Gaofeng Wang, and Hai Lin. "Modeling of Crosstalk Effects in Multiwall Carbon Nanotube Interconnects." *IEEE Transactions on Electromagnetic Compatibility* 54, no. 1 (2012): 133–139.

Newton, A. Richard and Takayasu Sakurai. "Alpha Power Law MOSFET Model and Its Applications to CMOS Inverter Delay and Other Formulas," *IEEE Journal Solid State Circuits* 25, no. 2 (1990): 584–594.

Orlandi, Antonio and Clayton R. Paul. "FDTD Analysis of Lossy, Multiconductor Transmission Lines Terminated in Arbitrary Loads." *IEEE Transactions on Electromagnetic Compatibility* 38, no. 3 (1996): 388–399.

Paul, Clayton R. "Incorporation of Terminal Constraints in the FDTD Analysis of Transmission Lines." *IEEE Transactions on Electromagnetic Compatibility* 36, no. 2 (1994): 85–91.

Paul, Clayton R. "Decoupling the Multi Conductor Transmission Line Equations." *IEEE* 44, no. 8 (1996): 1429–1440.

Pu, Shao-Ning, Wen-Yan Yin, Jun-Fa Mao, and Qing H Liu. "Crosstalk Prediction of Single- and Double-Walled Carbon-Nanotube (SWCNT/DWCNT) Bundle Interconnects." *IEEE Transactions on Electron Devices* 56, no. 4 (2009): 560–568.

Qian, Libo, Yinshui Xia, and Ge Shi. "Study of Crosstalk Effect on the Propagation Characteristics of Coupled MLGNR Interconnects." *IEEE* 15, no. 5 (2016): 810–819.

Rossi, Daniele, Jos Manuel Cazeaux, Cecilia Metra, and Fabrizio Lombardi. "Modeling Crosstalk Effects in CNT Bus Architectures." *IEEE Transactions on Nanotechnology* 6, no. 2 (2009): 133–145.

Yee, Kane S. "Numerical Solution of Initial Boundary Value Problem Involving Maxwell's Equations in Isotropic Media." *IEEE* 14, no. 3 (1966): 302–307.

Index